HONEY BEE PATHOLOGY

HONEY BEE PATHOLOGY

Leslie Bailey

Rothamsted Experimental Station, UK

1981

A Subsidiary of Harcourt Brace Jovanovich, Publishers

London • New York • Toronto • Sydney • San Francisco

ACADEMIC PRESS INC. (LONDON) LTD.
24/28 Oval Road
London NW1

United States Edition published by
ACADEMIC PRESS INC.
111 Fifth Avenue
New York, New York 10003

British Library Cataloguing in Publication Data

Bailey, L.
Honey bee pathology.
1. Bees – Diseases
I. Title
638'.15 SF538

ISBN 0-12-073480-X

LCCCN 81-68981

Typeset in Singapore by Colset Private Ltd.
Printed in Great Britain

PREFACE

This book is based upon "Infectious Diseases of the Honey-bee" (Bailey, 1963a) and incorporates much that has been learned in recent years, including knowledge of diseases and pathogens that were previously unknown but have proved to be unexpectedly widespread and common.

Most writers have discussed bees with little or no regard for other insects. This is an artificial separation which, although reasonably based on human interests, has often led to unreasonable anthropomorphic attitudes about bees; especially about their diseases. It can be corrected to some extent by considering honey bee pathology in the context of insect pathology. This subject has become too extensive and diverse to be summarized readily, but an awareness of it can give perspective and scale to a detailed account of the pathology of bees. This may well modify in return some of the attitudes that prevail about insect pathology, many of which have often been strongly influenced by well-established but erroneous beliefs about the diseases of bees.

Although much has developed since 1963 in honey bee pathology the treatment in this book is selective, for the sake of brevity and convenience; and it is based on experiments and observations that seem to be sound. Whenever possible, references are given to review and comprehensive articles, where further references can be found to special points.

Some knowledge of biology on the part of the reader is assumed, but, for those who are unfamiliar with biological terms, inexpensive scientific and biological dictionaries should be adequate.

Advanced accounts of the anatomy of bees are given by Snodgrass (1956) and Dade (1962). Wigglesworth (1972) and Roeder (1953) include much information about bees in their works on insect physiology.

I am indebted to many friends and colleagues, both scientists and beekeepers, at home and abroad, for their helpful and stimulating discussions. In particular, I thank Brenda V. Ball, for help with much recent research and with the manuscript, and Lynda Castle and J. Philip Spradbery, for many illustrations.

August 1981 *Leslie Bailey*

CONTENTS

Our postman said . . . "Isle of Wight disease? Never heard of it. My bees? No, I never lost none. John Preachy's? Why, of course they died; he used to feed 'em on syrup and faked-up stuff all winter . . . You can't do just as you like with bees. They be wonderful chancy things; you can't ever get to the bottom of they."

Adrian Bell (*The Cherry Tree*)

1. INTRODUCTION

Man has concerned himself about the diseases of honey bees for thousands of years. Aristotle (384–322 B.C.) described certain disorders, and Virgil and Pliny referred to some about the beginning of the first millennium. None of their descriptions is sufficient to identify the disorders with certainty. However, they made it plain that bees then were much the same as now and that the diseases we today call foulbrood and dysentery probably existed in antiquity. One description by Aristotle of a disorder of adult bees corresponds with that of one of the syndromes of paralysis (Chapter 3, I.A.).

In the more recent past, Shirach in 1771 described "Faux Couvain" (Steinhaus, 1956), which may well have been American or European foulbrood; and Kirby and Spence (1826) described "dysentery". Soon afterwards occurred one of the most significant events in insect pathology, and one that greatly influenced the concept of infectious diseases of all kinds, including those of bees. This was the demonstration by Louis Pasteur, in the mid-nineteenth century, of the way to rid the silkworm, *Bombyx mori*, of "pebrine"; a disease that was crippling the prosperous silk industry of France. He and his colleagues recognized the pathogen, which was later named *Nosema bombycis*, observed that it was transmitted in the eggs from infected females and, by microscopically examining the progeny of quarantined females for spores of the pathogen, were able to select healthy stocks and re-establish productive silkworm nurseries. Pasteur was greatly honoured by the silk industry and the French government for his classic solution of their problem. He, and others strongly influenced by him, went on from this success to establish the basic principles of infectious diseases of man and his domesticated animals. All kinds of severe diseases soon were found to be due to micro-organisms or viruses and the hunt for these became the dominant feature of disease investigations.

Great hopes and expectations also arose then about the diagnosis and cure of bee diseases. Dzierzon (1882) recognized that there were two kinds of foulbrood of bees: "mild and curable" of unsealed brood (undoubtedly European foulbrood), and "malignant and incurable" of sealed brood (clearly American foulbrood). Microbiological investigations into them were begun by Cheshire and Cheyne (1885). Entomologists also became impressed by the idea of spreading pathogenic micro-organisms among pest insects, hoping to control them with diseases as destructive as that which had ravaged the French silk industry and as those believed to be rife among bees.

The parasites that were newly found in sick bees quickly led to a common belief that bees suffered from a wide range of infections of great severity. It also became widely believed that the presence, or absence, of serious infectious disease was a matter of the presence or absence of the pathogen. When the pathogen was present, disease and eventual disaster were thought to be certain, as shown with pebrine in the silkworm and with several severe infectious diseases of other domesticated animals and of man.

The hopes and fears engendered about bees in those early days still continue but, as will be seen, most infectious diseases of bees are more subtle than previously imagined; and they are less amenable to treatment than pebrine, which had seemed such a formidable problem to Pasteur when he began his work.

2. THE HONEY BEE

1. NATURAL HISTORY

The honey bee colony has frequently been regarded either as an ideal society or as a kind of totalitarian state: it is neither. Social insects, whether termites (Isoptera), wasps, ants or bees (Hymenoptera), do not form organizations analogous to those of human societies. Their colonies are no more than families, often very large ones, but usually comprising one long-lived fertile female and her progeny; and each family is an independent unit which needs no contact with others apart from the occasional pairing of sexual individuals. Regarded in this way, social insects are not very different from the several million other known species of insects with which they form an intrinsically uniform group, especially with regard to their fundamental structure, physiology and pathology.

However, notwithstanding their close relationship with other insects, including some 10 000 species of bees of which about 500 are social, two of the four species of the genus *Apis*, the true honey bees, are sufficiently distinct to have long attracted the special attention of man. These are the European honey bee, *Apis mellifera*, and the very similar but physically smaller and quite distinct species, the eastern honey bee, *Apis cerana*. These two honey bee species have long been of particular interest to man because they store large amounts of accessible honey and can be induced to nest in movable containers or "hives". During the past few hundred years, the European honey bee has been taken by man all over the world and with particular success to the Americas, Australia and New Zealand. There are also several strains of *Apis mellifera* naturally distributed throughout the African continent. The eastern hive bee is restricted to S.E. Asia, China, east U.S.S.R. and Japan, and is to some extent being replaced by *Apis mellifera*, particularly in the temperate zones of these regions, by the activity of beekeepers.

A colony of honey bees is headed by a single queen and is composed of about 50 000 individuals on average. Worker bees clean and make the wax combs and feed brood in their first week or so of life, and then begin to forage, usually when they are 2 or more weeks old, first for pollen and then for nectar. They live no more than 4 or 5 weeks in summer, but in autumn, when nectar-flows and brood-rearing end, they hibernate as a cluster and individuals of the cluster may survive as long as 7 months. There are usually a few hundred drones in colonies in summer whose sole function is to mate with virgin queens. They are ejected from the colony by worker bees in autumn before the winter cluster forms.

Colonies reproduce by swarming. This usually means that the queen leaves the colony in early summer, attended by many, possibly more than half, of the workers, and goes to another suitable nest-site. The queenless colony that remains rears further queens, the larvae of which are usually being prepared at the time the swarm leaves. The first of these new queens to emerge usually kills the others before they emerge and thus becomes the new reigning queen.

When by any chance a colony loses its queen, a new one is reared from a young larvae which would otherwise have become a worker; but it is not known how a worker larva changes its development to become a queen.

The larval worker bee passes through the following six distinct phases in its life (Fig. 1):

1. The embryo develops for 3 days in the egg, which is fixed to the base of an open cell in the comb.
2. When the larva hatches from the egg it is fed continuously for the next 5 days, while it is growing in the open cell, by young adult bees or "nurse bees". The larva sheds its skin about every 24 h. The mid-gut of a growing larva is a blind sac (Fig. 2).
3. The fully grown larva is sealed in its cell by nurse bees and then spins a cocoon. This is discharged as a fluid from an orifice on its labium-hypopharynx or "lower-lip", and smeared over the cell walls where it becomes dry, tough and papery. At the same time the larvae discharges its faeces via the rectum, which temporarily joins up with the mid-gut for this purpose. The faeces become sandwiched between layers of the

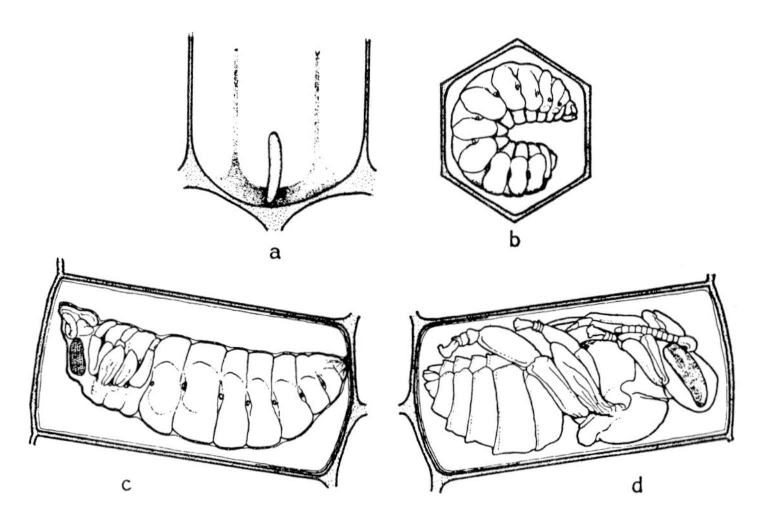

Fig. 1 The stages of development of a honey bee: (a) egg on the base of a cell in the comb; (b) larva about 4 days old in its open cell; (c) propupa and (d) pupa in their capped cells.

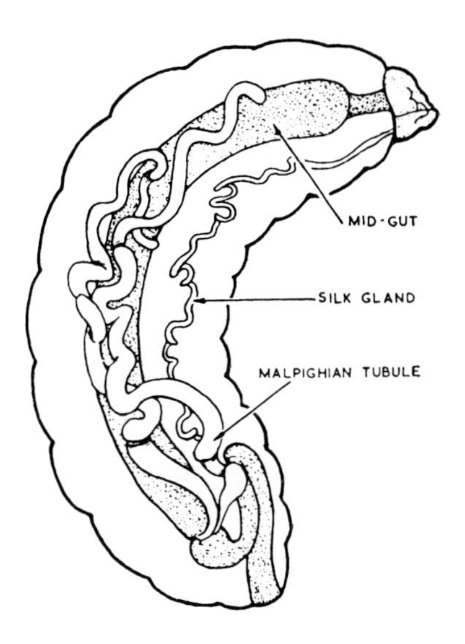

Fig. 2 Anatomy of the young larval honey bee. The mid-gut, hind-gut and Malpighian tubules are blind at their junction at this stage. (After Nelson, 1924).

cocoon. About 2 days after it is sealed over, the larva lies on its back with its head towards the cell capping.

4. The quiescent larva changes within a loosened fifth skin to a propupa, and after 2 days of this phase it sheds the fifth skin to become a white pupa.
5. The pupa, now resembling an adult bee in shape, slowly darkens in colour, beginning with the eyes.
6. The pupa sheds its skin, and a few hours later the adult insect emerges from its cell.

The pupal stage is shortest for the reproductive caste, "queen" and longest for the male, "drone". Queens emerge from their cells about 16 days after the egg is laid; the worker bees, which are genetically similar to queens but have undeveloped ovaries as well as other morphological differences, take about 21 days; and drones, which are haploid individuals, take about 24 days to develop. Drone larvae stay unsealed for about 2 days longer than worker larvae.

The adult bee eats pollen and honey, the latter being floral nectar concentrated by evaporation and with its sucrose content inverted by enzymes from the hypopharyngeal glands of adult bees until it is virtually an aqueous solution of about 30% glucose, 40% fructose, 8% maltose and other disaccharides, 2% sucrose and 0·5% organic acids. Pollen supplies all the protein fraction of the food and is eaten mainly by newly emerged and young adult bees in summer. The pollen is ingested into the crop in suspension in honey, from which it is separated, together with other particles, including those as small as bacteria, and passed into the mid-gut by the proventriculus. It is digested and absorbed by the gut and much of it is converted to a secretion of the hypopharyngeal glands of the head, from which it is ejected via the mouth as nitrogenous food for unsealed larvae, the adult queen and possibly for adult drones. Drones and queens are also able to feed themselves on honey, and drones probably feed themselves entirely in this way after their first few days or so of life. In autumn, when brood-rearing is almost over, protein is stored in the fat-body of adult bees as well as in the hypopharyngeal glands (Fig. 3). This reserve of protein probably helps the now rather inactive adult bees to survive the prolonged winter of temperate and sub-arctic climates and to have ready supplies of hypopharyngeal gland secretion for early spring brood-rearing.

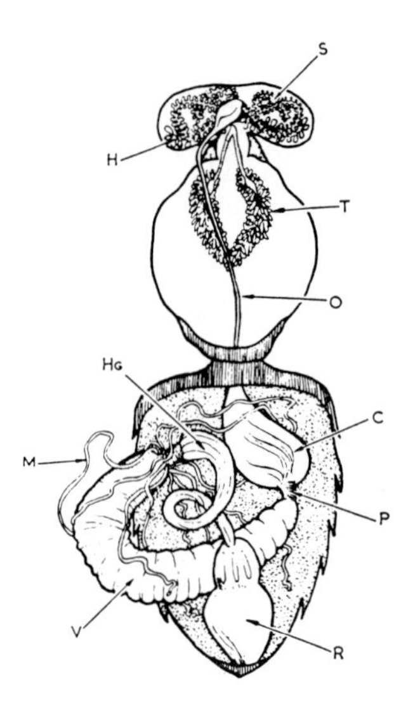

Fig. 3 Glands and viscera of the adult bee:
C = crop, H = hypopharyngeal glands, Hg = hind-gut, M = Malpighian tubules, O = oesophagus, P = proventriculus, R = rectum, S = head labial glands, T = thoracic labial glands, V = ventriculus (mid-gut). (After Snodgrass, 1956)

Larval food may be a mixture of secretions from several different glands of the adult bee, but there is little doubt that most of the protein, which comprises 40–60% of the dry matter of larval food, is from the hypopharyngeal glands. Carbohydrate, which forms 30–50% of the dry matter of larval food, is probably entirely from honey: it may form a larger proportion of the food of older larvae but although generally believed, this remains to be proved. Pollen accumulates in the gut of the larvae, but the amount is insignificant compared with the nitrogenous needs of the growing insect and its presence is probably fortuitous. Larval food like honey, is acid, the usual pH being about 4·0; 5–20% of the dry weight of larval food is fatty material. Much of this is 10 – hydroxydecenoic acid which is bactericidal at the normal pH of the food and comes from the mandibular glands.

II. BEEKEEPING

The honey bee evolved to the state in which we know it today long before the advent of mammals, not to mention man. Yet it is a popular belief among many biologists as well as beekeepers that bees are domesticated. The only insect that has been domesticated is the silkworm, *Bombyx mori*, which needs the care and attention of man in order to survive. By contrast, honey bees are feral insects no less than any of the millions of other insect species living in the forests, countryside and gardens. Honey bees can and do survive independently of man. Indeed, they must be left at liberty, even when in the hives of beekeepers, in order to survive. We have not learned how to keep them isolated, even partially, from their environment, whereas many species of wild animals, including a great variety of insects, can be readily propagated in entirely artificial conditions. Even if bees could be kept under such conditions, it would be of only academic interest because they would still have to be allowed to rove freely in order to collect nectar and to pollinate plants that need them. Beekeeping today is still as it has always been: the exploitation of colonies of a wild insect; the best beekeeping is the ability to exploit them and at the same time to interfere as little as possible with their natural propensities. The most productive strains of honey bee presently available for man are those that would survive best independently of him, because they are the ones that find and store most food.

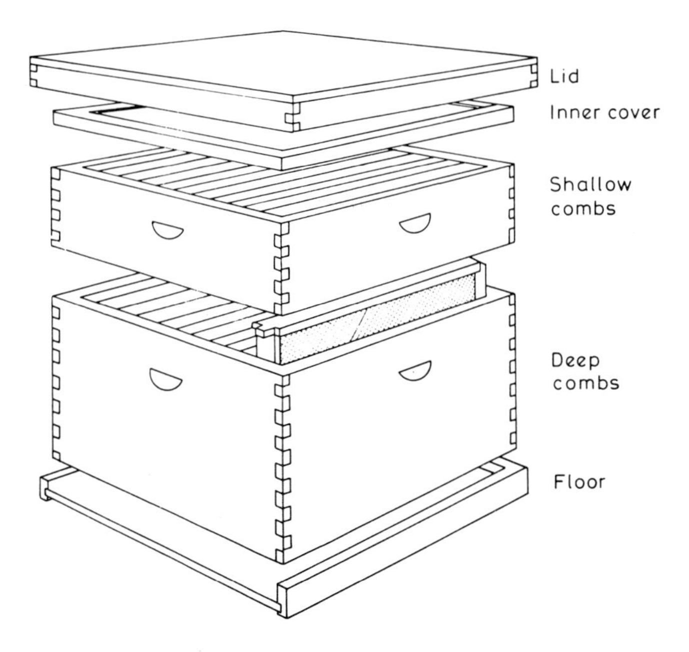

Fig. 4 A modern beehive.

Beyond providing a colony of bees with a weather-proof cavity of adequate volume in regions of abundant and varied nectar-yielding plants, the modern beekeeper can do little that is beneficial for his bees; although he can readily do a great deal that is harmful to them. All the methods and paraphernalia of beekeeping are entirely for his convenience. Bee colonies can live successfully and indefinitely in a suitably sized cavity of no particular shape as well as in any beehive. Bees will successfully occupy hollow logs, drain pipes, baskets and more unlikely containers, as has been well known to beekeepers for millenia. All the refinements have come from the wish to remove honey easily, with least harm to the bees, from colonies kept in readily transportable hives.

The ultimate achievement has been to make rectangular frames, usually of wood, in each of which bees will readily build one of their naturally orderly vertical combs (Fig. 27a). These frames are hung in a box, with a space of about 7–9 mm between the combs, and between the ends and top of the frames and the sides and top of the box. The bees accept this space as that of a thoroughfare and so do not usually block it up with wax and propolis, the way they quickly block narrower or wider gaps. The beekeeper can then easily remove, replace or rearrange the frames without much harm to the bees, and can extract the honey from the comb, usually in a special kind of centrifuge. This causes little harm to the combs, which are the items most valuable to the beekeeper and which can be returned to the hive for the bees to use again and again. Every significant feature of the several different kinds of successful modern beehive, however simple or complicated their wooden structure may be, is based on the existence of the bee space, which was first recognized by Langstroth in America in 1851.

Modern beehives (Fig. 4) are rectangular boxes of combs that have a loose lid and stand on loose floors. Each floor is constructed to form a narrow horizontal gap below the edge of the bottom box to form the entrance. Whole hives of this construction can easily be strapped up and stacked for transportation, and boxes of comb are simply piled one on another to make room as required for growing colonies and stored honey.

3. VIRUSES

Viruses are little more than genetic material enclosed in a protein shell or coat. They do not possess the mechanisms that would enable them to multiply independently by assimilating nutrients in the manner of most micro-organisms, such as bacteria; they can multiply only within the living cells of their host. When a virus infects a cell, it uses the

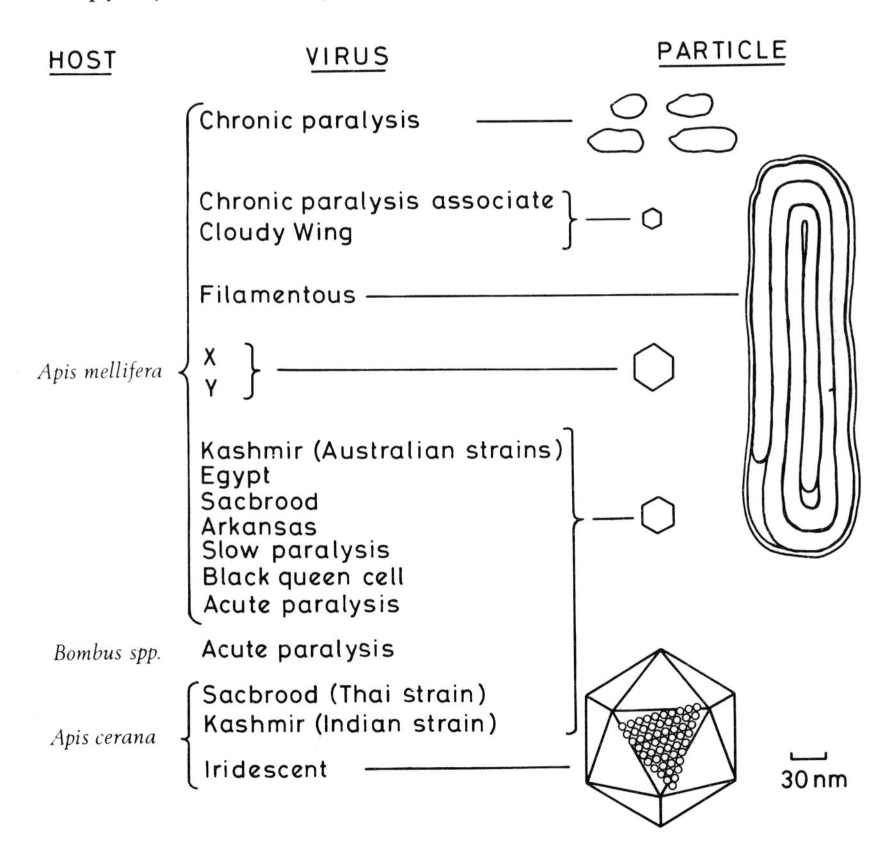

Fig. 5 List and diagrammatic outlines of particles of viruses that attack honey bees.

TABLE I. Properties of honey bee viruses*

Virus	Dimensions (nm)	$S_{20,w}$ (Svedbergs)	Buoyant Density in CsCl (g/ml)	Nucleic Acid		Proteins
				Type	Molecular wts ($\times 10^{-6}$)	Molecular wts ($\times 10^{-3}$)
Chronic paralysis	20 × 30 to 60	80 to 130	1·33	RNA	1·35, 0·9, 0·35**	23·5
Chronic paralysis virus associate	17	41	1·38	RNA	0·35**	15
Cloudy wing particle	17	49	1·38	RNA	0·45	19
Acute paralysis	30	160	1·37	RNA		23,31
Arkansas	30	128	1·37	RNA	1·4, 1·8	41
Black queen cell	30	151	1·34	RNA	2·8	6,29,32,34
Egypt	30	165	1·37	RNA	N.D.	25,30,41
Kashmir (Indian strain: *Apis cerana*)	30	172	1·37	RNA	N.D.	24,37,41
Kashmir (Australian strains)	30	172	1·37	RNA	N.D.	25,33,36,40,44
Sacbrood	30	160	1·35	RNA	2·8	26,28,31
Sacbrood (Thai strain: *Apis cerana*)	30	160	1·35	RNA	2·8	30,34,39
Slow paralysis	30	176	1·37	RNA	N.D.	27,29,46
X	35	187	1·35	RNA	N.D.	52
Y	35	187	1·35	RNA	N.D.	50
Iridescent (*Apis cerana*)	150	2216	1·32	DNA	N.D.	N.D.
Filamentous	150 × 450	N.D.	1·28	DNA	12·0	13 to 70***

* Mostly from Bailey (1976), or original. ** Hilary Overton, Imperial College (unpublished data). *** About 12 proteins. N.D. = not determined.

cellular apparatus to make copies of itself. This can continue, without much obvious change to the cell, as long as the organism of which the cell is a part remains alive; but usually, infected cells become damaged, die and disintegrate, thereby releasing very many infective virus particles. These particles, or virions, are minute and usually far too small to be seen by light microscopy.

All forms of life are attacked by viruses, and insects of all kinds become infected by a wide variety of virus types. These are usually host-specific, or have a very limited host-range, and the virions of several different kinds of well-known insect viruses become embedded in crystalline matrices of protein, ''polyhedra'', which are usually large enough to be seen easily by light microscopy. These embedded viruses are peculiar to insects, mostly to the larvae of Lepidoptera (Fig. 32c), and there are very many known examples. Comparatively few viruses that have small non-embedded virions resembling the kinds that attack most other animals and plants, have so far been identified in insects. Only about 30 or 40 different ones are known and half or more of them have been found in the honey bee (Figs 5, 28; Table I).

I. HONEY BEE VIRUSES THAT OCCUR IN BRITAIN AND ELSEWHERE

A. Paralysis

1. Symptoms

This virus disease has two distinct sets of symptoms, or syndromes (Bailey, 1975). One of these (Type 1), seemingly the commonest in Britain and described by beekeepers as ''paralysis'' more than a 100 years ago, includes an abnormal trembling motion of the wings and bodies of affected bees. These fail to fly but often crawl on the ground and up grass stems, sometimes in masses of thousands of individuals. Frequently they huddle together on top of the cluster in the hive. They often have bloated abdomens and partially spread, dislocated wings (Fig. 29b).

The bloated abdomen is caused by distension of the honey sac with liquid (Fig. 29d). The mechanical effect of this accelerates the onset of so-called ''dysentery'', and sick individuals die within a few days. This kind of disease seems to correspond to the disease long known in Europe as *Waldtrachkrankheit*, so named because it often seems to be associated with nectar gathered from the forests.

The other syndrome (Type 2) has been given a variety of names: ''black robbers'' and ''little blacks'' in Britain, *Schwarzsucht* and *mal noir* or *mal nero* in continental Europe; and could well have been the condition described by Aristotle of a black bee with a broad abdomen which he called ''a thief'' (ϕ *ẃρ*). At first the affected bees can fly, but they become almost hairless, appearing dark or almost black which makes them seem smaller than usual but with a relatively broad abdomen; and they are shiny, appearing greasy in bright light (Fig. 29c). They suffer nibbling attacks by older bees in the colony, which may account for their hairlessness and, when they fly, are hindered from returning to their colony by the guard bees, which makes them seem like robber bees. In a few days they become trembly and flightless and soon die. Both syndromes often occur in one colony, but usually one or the other predominates.

Sections of the hind-gut epithelium of paralytic bees show basophilic cytoplasmic bodies (Fig. 29f), which seem to be specific to the disease and were first described by Morison (1936) who suspected they were associated with a virus.

2. Cause

The virus that causes paralysis (Figs 5, 28b), is called chronic paralysis virus to distinguish it from acute paralysis virus (Section G) which was found at the same time (Bailey, 1976). The properties of chronic paralysis virus particles are given in Table I. When injected into, or, fed to, or sprayed on adult bees, purified preparations of the particles cause paralysis, usually with the Type 1 syndrome, and the difference between the syndromes probably expresses genetical differences between individual bees. There is considerable evidence that susceptibility to the multiplication of chronic paralysis virus is closely limited by several inherited qualities of bees and some variation of these qualities might well lead to variations in the symptoms. Rinderer *et al.* (1975) and Kulincevic and Rothenbuhler (1975) were able to select strains of bees which were more susceptible than usual to a ''hairless black syndrome'', later shown to be chronic paralysis by Rinderer and Green (1976). Other circumstantial evidence indicating that susceptibility to paralysis is closely limited by hereditary factors has been discussed by Bailey (1967d, 1965a). Inbreeding with colonies that have paralysis, or allowing them to rear their own queens that mate with drones from similar colonies maintains a higher incidence of the disease in them than when they are supplied with queens from elsewhere.

3. Occurrence

Chronic paralysis virus has been detected serologically in extracts of bees found with paralysis symptoms in Australia, New Zealand, China, Mexico, U.S.A., Scandinavia, continental Europe, the Mediterranean area and many parts of Britain; and virus particles with the same appearance have been described occurring in the Ukraine, France and Canada. Infectivity tests with extracts of bees from apparently normal colonies in Britain have shown that the virus is commonly distributed among them throughout the year and causes mortality that sometimes approaches 30% of the total usually accepted as normal. Outbreaks of severe disease are erratic and follow no seasonal pattern (Bailey, 1976; Bailey *et al.*, 1980b).

4. Multiplication and spread

Many millions of particles of chronic paralysis virus can be extracted from one bee with paralysis. Many tissues become infected with the virus, including the brain and nerve ganglia. Occasionally pupae are killed by the virus at a late stage in their development in colonies suffering severely from paralysis.

Very many millions of particles are needed to infect a bee by mouth and cause paralysis, but about 100 or fewer will cause the disease when injected into the haemolymph. However, a likely method of infection in nature, which requires only few particles, is via pores in the cuticle left by broken bristles. This briefly exposes the cytoplasm of epithelial tissue and when bees are crowded together some virus can become rubbed into the wound.

Much chronic paralysis virus is in the distended honey-sacs of paralytic bees and in the pollen collected by apparently normal individuals of colonies suffering from paralysis. The virus is probably secreted by the bees from their food glands into the liquid that enters the honey-sac and that they add to the pollen they collect (Bailey, 1976). Perhaps of greater significance is the fact that chronic paralysis virus occurs commonly in colonies that are accepted by beekeepers as healthy. Sensitive infectivity tests have shown that apparently normal live bees often contain some of the virus. There is no particular time of year when paralysis, or the virus in seemingly healthy colonies becomes most common. Therefore, chance rather than seasonal events may determine the rate at which it spreads between bees. The unusual crowding of bees within the colony, which may occur for a variety of reasons, both natural and artificial, and a consequent increase of transmission of virus via the pores left in the cuticle by broken bristles, would be compatible with the irregularities of infection and of outbreaks of paralysis. Kulincevic *et al.* (1973) observed that symptoms of paralysis occurred sooner in bees when they were deprived of their queen, but the reason is unknown.

B. Chronic Bee-paralysis Virus Associate

A virus-like particle, 17 nm across (Figs 5, 28a) is consistently associated with chronic bee-paralysis virus but is serologically unrelated to this virus. It does not multiply when injected alone into bees, and therefore may be a satellite of the paralysis virus, depending on it in the way that similar small particles occurring in plants and animals need genetic information supplied by other viruses in order to replicate. As with these satellites and their helper-viruses the associate particle interferes with the multiplication of chronic paralysis virus in individual bees, inhibiting particularly the relative amount of the longest, most infective particles. It is more evident in queens than in workers (Bailey *et al.*, 1980a) and may be of some significance in, or a reflection of, the defence mechanisms of individuals against paralysis.

C. Sacbrood

1. *Symptoms*

Whereas healthy bee larvae pupate 4 days after they have been sealed in their cells, larvae with sacbrood fail to pupate, and remain stretched on their backs with their heads toward the cell capping. Fluid then accumulates between the body of a diseased larva and its tough unshed skin, and the body colour of the larva changes from pearly white to a pale yellow (Fig. 29). Finally, after it has died a few days later, it becomes dark brown. The head and thoracic regions darken first and at this stage (Fig. 29i) the signs are most distinctive and specific for sacbrood. Finally, the larva dries down to a flattened gondola-shaped scale.

2. Cause

The properties of sacbrood virus particles are given in Table I. When added to the food of unsealed larvae in bee colonies, the larvae die of sacbrood shortly after they have been sealed in their cells. Larvae about 2 days old are most susceptible.

3. Occurrence

Sacbrood probably occurs throughout the world. The disease was first identified by White (1917) in the USA and shown by him to be caused by a filterable agent. The virus was identified by Bailey *et al.*, (1963) and has been detected in bees sent to Rothamsted from Europe, Egypt, Australia, New Zealand and New Guinea. Sacbrood is extremely common in Britain, although it was believed not to occur there until the virus was identified in 1963 (Bailey, 1975). Before then the disease was believed to be a non-infectious hereditary fault known as "addled brood", mainly because experimenters had failed to spread the disease by placing combs containing diseased larvae in healthy colonies. However, sacbrood does not spread readily this way (Section C4). Recent surveys in England and Wales show that most colonies are infected and, although most show no signs, up to 30% usually contain a few larvae with sacbrood.

A strain of sacbrood virus has been isolated from larvae of *Apis cerana* from Thailand. It is closely related to sacbrood virus of *Apis mellifera*, but has distinctive properties (Tables I, VII; Bailey *et al.*, 1981).

4. Multiplication and spread

Sacbrood virus multiplies in several body tissues of young larvae but they continue to appear normal until after they are sealed in their cells. Then they are unable to shed their last larval skin, because the thick tough endocuticle remains undissolved, and they die. Presumably, infection prevents the usual formation of chitinase by damaging the dermal glands. Each larva killed by sacbrood contains about a milligram of sacbrood virus, enough to infect every larva in more than a 1000 colonies. Yet in natural circumstances, sacbrood usually remains slight, and usually abates markedly and spontaneously during the late summer. This is because adult bees detect many larvae in the early stages of sacbrood and remove them from the bee colony, and because the virus quickly loses infectivity in the dried remains of those that are left.

Continuity of infection from year to year is provided by adult bees in which sacbrood virus multiplies without causing obvious disease. The youngest workers are the most susceptible and probably become infected in nature mostly when they remove larvae killed by sacbrood. During this activity, they ingest liquid constituents, especially the ecdysial fluid, of larvae that become damaged in the process. Within a day after young bees ingest such material, much sacbrood virus begins to collect in their hypopharyngeal glands (Bailey, 1969a). Infected nurse bees probably transmit sacbrood virus when they feed larvae with secretions from these glands.

However, infected adult bees cannot be very efficient vectors, or else they must usually be prevented from transmitting much virus, otherwise sacbrood would not subside spontaneously in summer. Much evidence shows that bees are usually prevented from transmitting the virus by behavioural changes (Bailey and Fernando, 1972). Infected young bees cease to eat pollen and soon cease to feed and tend larvae. They fly and forage, but do so much earlier in life than usual, and they almost all fail to collect pollen (Table II). The few that do collect pollen add much sacbrood virus to their pollen loads, probably in the gland secretions they add to pollen as they collect it. Were many infected bees to tend larvae and later gather pollen, which is quickly consumed by young susceptible individuals, much virus would soon reach and kill more larvae. Sacbrood virus put into nectar gathered by infected bees is a far less important source of infection because the incoming nectar is much diluted among the rest and is quickly and widely distributed within the bee colony, whereas pollen loads remain intact within the cell of the comb where they are placed. Any virus in them remains concentrated and more likely than virus in nectar to infect a bee. Times when the division of labour of bees is least well developed, such as the early part of the year or prolonged periods of dearth, are when sacbrood virus is likely to be transmitted from infected adults to larvae.

TABLE II. Numbers of marked bees seen foraging, and (in parenthesis) the percentage collecting pollen, after equal numbers were either infected with sacbrood virus or were untreated when 5 days old (after Bailey and Fernando, 1972)

	Treatment	
Days after infection	*Infected*	*Nil*
1–5	64 (3·1)	0
6–10	140 (0·7)	58 (3·4)
11–15	14 (0)	229 (17·0)
16–20	0	46 (10·8)
21–25	0	0

Interestingly, exactly the same permanent changes in behaviour occur in young worker bees that are briefly anaesthetized with CO_2, or other forms of anoxia (Ribbands, 1953), as occur in those infected with sacbrood virus, including a permanent loss of appetite for pollen (Bailey, 1969a). They are equivalent to the changes that occur with age in healthy bees, and the same mechanism may be activated by sacbrood virus and by CO_2. It may well be a response to acidosis either caused by CO_2 or following damage to oxidative processes in tissues caused by ageing or by sacbrood virus.

Accompanying the behavioural changes, the metabolic rate of infected workers is diminished and their lives are shortened to about the same length as healthy workers that are completely deprived of pollen. These effects of sacbrood virus further decrease its chances of spread and of surviving the winter when infected bees are most likely to become chilled and lost from the cluster. The lives of drones, which do not eat pollen, are seemingly unaffected by the virus, although remarkable quantities of sacbrood virus multiply in their brains.

D. Black Queen Cell Virus, Filamentous Virus and Bee Virus Y

These are three common viruses of special interest because they are intimately associated with *Nosema apis* (Bailey *et al.*, 1980b).

Black queen cell virus, as its name suggests, is associated with queen cells that develop dark brown to black cell walls. They contain dead propupae or pupae in which are very many particles of the virus. In the early stages, infected pupae have a pale yellow appearance and a tough sac-like skin, resembling propupae that have died of sacbrood. They are most noticeable when many queen cells are being reared together in "queen rearing" (broodless and queenless) colonies from young larvae that have been "grafted" or inserted into the colonies by the usual techniques of beekeepers (Laidlaw, 1979), and they are most usual in the early part of the season. By contrast with sacbrood virus, black queen cell virus does not readily multiply when ingested by young worker larvae, young worker bees or drones, nor when injected into adult workers or drones (Bailey and Woods, 1977) but it is a common infection of adult bees in the field (Fig. 6).

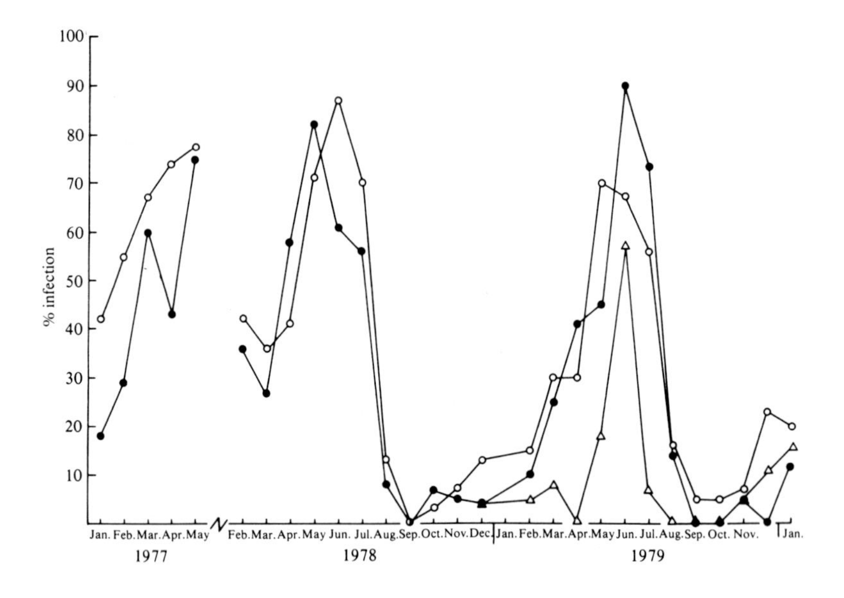

Fig. 6 Mean percentage of 25 undisturbed bee colonies infected with black queen cell virus (●) and bee virus Y (△, not identifiable until December 1978) and of individuals in them infected with *Nosema apis* (o). (From Bailey *et al.*, 1980b).

Filamentous virus (Figs 5, 28), which was originally believed to be a rickettsia (Wille, 1967), was identified as a virus in the U.S.A. by Clark (1978). It multiplies in fat-body and ovarian tissues of adult bees and the haemolymph of severely infected bees becomes milky-white with particles, but no other symptoms are known.

Bee virus Y (Figs 5, 28), occurs frequently in adult bees during the early summer (Fig. 6). Experimentally, it has been found to multiply only when given in food to adult bees but there are no known symptoms.

TABLE III. Comparative incidence of *Nosema apis* and viruses in 175 bee colonies (From Bailey *et al.*, 1980b)

Virus	No. Colonies with:	
	N. apis + virus	Virus only
Black queen cell	46	1
Bee virus Y	38	0

All three of these viruses almost invariably multiply only in those individual adult bees that are also infected with *Nosema apis* (Bailey *et al.*, 1980b; Bailey, 1981) (Tables III and IV). The viruses are unrelated to each other, and seem unlikely to have a common relationship with *N. apis*. A possible reason for their close association with the parasite is that *N. apis* decreases the resistance of bees to infection by viruses that usually invade via the gut. Infection by *N. apis*, which multiplies within the cytoplasm of the cells of the mid-gut epithelium, could be expected to interfere with or prevent the production of a resistance factor, such as the anti-viral protein produced by the gut of silkworms (Hayashiya *et al.*, 1969).

TABLE IV. Distribution of viruses within dead field bees after separating them into those with and without *Nosema apis* (Original data).*

Virus	No. Bees Examined	Groups or Individuals with:	
		N. apis + virus	Virus only
Filamentous	32 groups of 10	13	0
Black queen cell	6 groups of 30	4	0
Y	6 groups of 30	3	0
X	452 individuals	28	73

* Bees were from colonies already known to be infected with *N. apis* and one or other virus.

The viruses presumably add to the pathogenic effect of *Nosema apis*, and their presence or absence may account for the considerable variations of virulence that have been attributed to the microsporidian.

Black queen cell virus, bee virus Y, and filamentous virus have been identified in bees sent to Rothamsted from Britain, N. America and Australia. Filamentous virus has also been found in bees sent from Japan and Batuev (1980) has reported it in the USSR.

E. Bee Virus X

This virus (Figs 5, 28), distantly related serologically to bee virus Y (Bailey *et al.*, 1980c), also occurs in adult bees and, again resembling bee virus Y, has been found to multiply experimentally only when given in food to adult bees. Yet it has no relationship with *Nosema apis* (Table IV), is less common than bee virus Y, and is prevalent at a different time of year to bee virus Y (Bailey *et al*, 1980b). Bee virus X shortens the lives of bees significantly when given to them in food.

Recent evidence shows that bee virus X is associated with *Malpighamoeba mellificae* (Chapter 6, II.) in late winter more frequently than can be expected by chance (Bailey, 1981). This probably means that the virus is transmitted by severe faecal contamination, in the same manner as *M. mellificae* (Chapter 6, II.C.); but additional factors must affect the issue, otherwise bee virus X would also tend to occur with bee virus Y and *Nosema apis* (Section D.).

F. Cloudy Wing Particle

This is a common virus of bees which sometimes show a marked loss of transparency of their wings when they are severely infected. The particles are 17 nm across (Figs 5, 28) and observations suggest infection is airborne between bees over a short distance (Bailey *et al.*, 1980a). Crystalline arrays of the particles occur between the muscle fibrils (Fig. 32b) to which they may be conducted via the tracheal system. Infected individuals soon die.

About 15% of colonies have been found to be infected by the virus in Britain (Bailey *et al.*, 1980b) and the virus has also been detected in samples of bees sent to Rothamsted from Egypt and Australia.

G. Acute Bee-paralysis Virus

This virus was discovered as a laboratory phenomen during work on chronic bee-paralysis virus (Bailey *et al.*, 1963). Extracts of chronically paralysed, or of seemingly healthy bees, were injected into further healthy bees and some of these were killed by acute paralysis virus. Further investigations showed that this virus occurs commonly in seemingly healthy bees in Britain, especially during the active season (Fig. 7), but, again in Britain, it has never been associated with disease or mortality of bees in nature. It has been identified at Rothamsted in bees sent from Moscow where it was said to be causing mortality in the field, and, according to Batuev, (1979), is transmissible by *Varroa jacobsoni* (Chapter 7, IV.). The virus has also been identified in bees sent to Rothamsted from Belize in amounts large enough to have been the cause of their death (Bailey *et al.*, 1979). Therefore, it

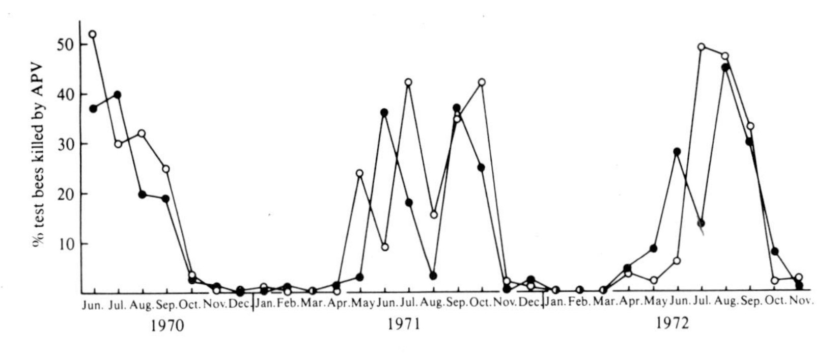

Fig. 7 Mean percentage of test bees killed by acute paralysis virus when injected with extracts each of 20 live seemingly healthy adults from each of 2 normal colonies at 1 site (●) and 2 or 3 colonies at another (o). (From Bailey *et al*, 1980b).

appears to kill bees in nature, but usually, if not always, remains suppressed in Britain. It does not readily infect bees other than by injection so may require a vector that pierces the body wall of adult bees, such as *V. jacobsoni*, to inject it into the haemolymph in which it is carried to tissues more vulnerable or more vital than those where it usually multiplies. However, there is no evidence that it causes the death of bees in Britain in spite of the common occurrence of *Acarapis woodi* (Chapter 7, I.) which might be expected to transmit it similarly. Usually, it appears to be contained within tissues that are not immediately essential to the life of the bee. It occurs in the pollen loads of foraging bees and in their thoracic salivary glands, and similarly in bumble-bees; but it was not found in the pollen of plants (*Trifolium pratense*) visited by pollen-collecting bumble-bees (Bailey, 1975), so it seems unlikely to be a plant virus.

H. Slow Paralysis Virus

This virus has been found occasionally in extracts of adult bees and causes their death about 12 days after it is injected into their body cavity. Nothing is known of its occurrence and no disease has been associated with it in nature (Bailey, 1976).

II. HONEY BEE VIRUSES THAT DO NOT OCCUR IN BRITAIN

A. Kashmir Bee Virus

This virus was isolated from diseased adults of *Apis cerana* sent to Rothamsted from Kashmir (Bailey and Woods, 1977) and from Mahableshwar, India (Bailey *et al.*, 1979). It was in company with *Apis* iridescent virus (Section D) in the Kashmir bees, but was alone and in large amounts in the bees from India. When injected into adults of *Apis mellifera*, or when rubbed on their bodies, it multiplies quickly and profusely and kills them within 3 days.

Strains of Kashmir bee virus have also been found in adults of *Apis mellifera* in Australia (Bailey *et al.*, 1979). Three Australian strains of the virus have been identified so far. All are closely related serologically, and they are somewhat less related to the type strain from Kashmir than they are to each other. The Australian strains were associated with severe mortality of adult bees in the field.

There is no evidence that Kashmir bee virus occurs in *Apis mellifera* anywhere other than in Australia, and its presence there was unexpected because *A.cerana* does not occur in the continent. Possibly, Kashmir bee virus in Australia has come from other insect species that are native to both Australia and south-east Asia, such as the ''sweat'' or ''stingless'' bees of the genus *Trigona*. Its occurrence in *Apis mellifera* in Australia is a rare example of the recent acquisition of a virus in nature by an insect. The instability of the proteins of the Australian strains of Kashmir bee virus and their serological differences (Bailey *et al.*, 1979) contrast with the stability and uniformity of the other known viruses of *Apis mellifera*. This may reflect a process of mutation and selection that is still being undergone by Kashmir bee virus as it adapts to the honey bee in Australia.

B. Arkansas Bee Virus

This virus was found as an inapparent infection of bees in Arkansas (Bailey and Woods, 1974) by injecting apparently healthy bees with extracts of pollen loads trapped from foragers returning to their colonies. The injected bees died after 15–25 days. It has recently been identified, together with chronic paralysis virus, in dead bees collected from dwindling colonies in California and sent to Rothamsted, but it has not been identified anywhere other than in the U.S.A..

C. Egypt Bee Virus

Egypt bee virus was isolated from adult bees (*Apis mellifera*) in Egypt (Bailey *et al.*, 1979). Nothing is known of its natural history or of its effects.

D. *Apis* Iridescent Virus

This is the only example of an iridovirus from Hymenoptera and was isolated from adults of *Apis cerana* sent to Rothamsted from Kashmir (Bailey *et al.*, 1976). Iridoviruses are so called because the crystalline masses they form when purified, or even in the tissues where they multiply, appear blue-violet or green when illuminated with bright white light. Many examples are known in a wide range of insects but none are related to *Apis* iridescent virus, although they are physically indistinguishable from it.

Apis iridescent virus is associated with ''clustering disease'' of *Apis cerana* in India. The most striking and consistent sign of this is an unusual inactivity, especially in summer, of colonies that frequently form small, detached clusters of flightless individuals and often lose many bees crawling on the ground. Large colonies perish within 2 months of becoming visible affected (Bailey and Ball, 1978).

Iridescent virus-infected tissues can be seen easily by microscopic examination, best of all in fresh unfixed specimens, but readily enough in bees that are preserved in formalin or alcohol. Infected tissues appear bright blue, in striking contrast with the surrounding creamy white tissues. Single infected cells can be distinguished in otherwise healthy tissue.

Apis iridescent virus has been found only in *Apis cerana* from Kashmir and Northern India, and may be limited in nature to this species in the Himalayan regions, although it can multiply in *Apis mellifera*.

Nothing is known of the natural history of the virus. It multiplies in many different tissues: fat body, alimentary tract, hypopharyngeal glands and ovaries; so it could be transmitted via faeces, eggs, gland secretions or even by ectoparasites. However, it is not transmitted by tracheal mites, such as *Acarapis woodi*, which were first suspected to be the

TABLE V. Cultivation of honey bee viruses

Virus	Instar*	Method of Infection**	Incubation period (days)***
Chronic paralysis (CPV)	A	I	7
	P	I	5
Chronic paralysis virus associate (CPVA)	A (queens)	I	7
	P (queens)	I	5
Cloudy wing particle (CWP)	A	?	?
Acute paralysis virus (APV)	A	I	5
	P	I	5
Arkansas bee virus (ABV)	A	I	21
	P	I	5
Black queen cell virus (BQCV)	P	I	5
Egypt bee virus (EBV)	P	I	5
Kashmir bee virus (KBV)	A	I	3
	P	I	3
Sacbrood virus (SBV)	L	F	7
	P	I	5
Sacbrood virus, Thai strain (TSBV)	P	I	5
Slow paralysis virus (SPV)	A	I	12
	P	I	5
Apis iridescent virus (AIV)	P	I	5
Bee virus X (BVX)	NA	F	30
Bee virus Y (BVY)	NA	F	30
Filamentous virus (FV)	A	?	?

* A = adults in cages at 35°C; P = pupae in Petri dishes at 96% R.H. and 35°C; L. = larvae 2 days old, kept in bee colonies until cells sealed, then incubated without adult bees at 35°C. NA = newly emerged adults in cages at 30°C supplied with pollen.

** I = by injection into haemocoele through abdominal intersegmental membrane; F = in food.

*** Based on infection with minimum infective doses.

cause of clustering disease, because these were rare or absent in many samples of sick bees, every individual of which was infected with *Apis* iridescent virus (Bailey and Ball, 1978). Curiously enough, the virus does not multiply when injected into larvae of the greater wax moth *Galleria mellonella*, whereas many of the iridescent viruses from other insects multiply readily in wax moths.

III. CULTIVATION AND PURIFICATION OF BEE VIRUSES

Many bee viruses can be cultivated in the laboratory, for experimental work and for the production of antisera, either in adult bees or in bee pupae (Table V).

Adult bees are best collected by placing a comb with adhering bees into a suitable box that can be filled with carbon dioxide. The gas is taken from a pressurized cylinder and passed through water, or a large empty vessel, to remove or melt frozen particles of solid carbon dioxide, which are very injurious to bees. The bees quickly become anaesthetized and can be placed in suitable cages (Fig. 27b). They are allowed to recover at about 20°C before they are incubated in a dry atmosphere at 30°C or 35°C. They are anaesthetized again when they are injected. This is best done the next day because bees sometimes soon die after they have been anaesthetized more than once in 24 h.

The methods found suitable for the purification of bee viruses are given in Tables VI and VII. These methods are neither unique nor inflexible, neither is it likely that they cannot be improved. More than one method is suitable for some of the viruses, but some methods are not suitable for certain viruses, as indicated.

Ammonium acetate buffer is better than phosphate for electron microscopy of preparations negatively stained with neutral phosphotungstate or ammonium molybdate, but phosphate buffer is better for serology. Immunodiffusion tests as described by Mansi (1958) can be done with purified virus or crude extracts made by grinding the head or

TABLE VI. Outline of method for extraction and purification of bee viruses

1. Grind insects in buffer and solvents. (a)
2. Filter through cotton or cheese-cloth.
3. Clarify by slow-speed centrifugation. (b)
4. Sediment virus from supernatant fluid by high-speed centrifugation. (c)
5. Resuspend pellet in buffer. (d)
6. Clarify by slow-speed centrifugation.
7. Repeat 4, 5 and 6 (except for (d)2 and (d)4 below).
8. Centrifuge down 10–40% (W/V) sucrose gradients. (e)
9. Locate and remove virus fractions.
10. Dialyse virus fractions against buffer.
11. Sediment virus by high-speed centrifugation.

a, b, c, d, e: see Table VII for details.

TABLE VII. Details of methods for extraction and purification of bee viruses

	Virus	Special Modifications and Additional Treatments
(a) Extraction fluid		
1. 0·01 M potassium phosphate (P), pH 7·0 + 0·02% diethyldithiocarbamate (DIECA) + 1 vol. diethyl ether (E) following by emulsification with 1 vol. carbon tetrachloride (CCl_4)	All except CWP TSBV and EBV	nil
	FV	no E or CCl_4
	AIV	no E
2. 0·1 M ammonium acetate (AA), pH 7·0 + 0·02% DIECA	FV	nil
3. 0·5 M P, pH 8·0 + 0·02% DIECA + 1 vol. E followed by emulsification with 1 vol. CCl_4	CWP, SBV	nil
	TSBV	+ 0·02 M sodium ethylene diaminetetra-acetate (EDTA)
	EBV	1% ascorbic acid + 2% EDTA instead of DIECA
(b) Slow speed centrifugation		
1. 8000 *g* for 10 min	All except FV and AIV	nil
2. 150 *g* for 10 min	FV, AIV	nil
(c) High speed centrifugation		
1. 75 000 *g*/3·0 h	All except CWP CPVA, FV and AIV	nil

TABLE VII (contd.)

	Virus	Special Modifications and Additional Treatments
2. 75 000 *g*/3·5 h	CWP, CPVA	nil
3. 30 000 *g*/30 min	FV, AIV	nil
(d) Treatment of sedimented virus		
1. Resuspend in 0·01 M P pH 7·0	All except FV, CWP, EBV, TSBV, and BVY	Keep extracts from pupae at 5°C for several hours to a few days and clarify by low-speed centrifugation, also at 5°C (not suitable for CPV)
2. Resuspend in 0·1 M AA, pH 7·0	FV	Layer on 50% w/v sucrose and sediment at 75 000 g/3 h; resuspend in 0·1 M AA. pH 7·0
3. Resuspend in 0·5 M P, pH 8·0	CWP, EBV	nil
	TSBV	+ 0·02 M EDTA
4. Resuspend in 0·01 M P, pH 7·0, add an equal volume of 0·2 M AA, pH 5·0 (not suitable for ABV, EBV or SBV; less suitable for yield but better for purification of CPV from pupae than 1.)	BVY, BVX BQCV, CPV APV	Clarify by low-speed centrifugation; sediment virus by high speed centrifugation; resuspend in 0·1 M AA, pH 7·0 or 0·1 M P, pH 8·0
(e) Sucrose gradients		
1. 45 000 *g*/3 h at 4°C	All except CPVA, CWP and FV	
2. 45 000 *g*/4·5 h at 15°C	CPVA, CWP	Sucrose in appropriate buffers, as in (d).
3. 10 000 *g*/30 min at 5°C	FV	

abdomen of a bee in 0·05 ml of 0·85% saline + 1 drop of diethyl ether in a small conical tube. The agar for the test contains 0·05 M potassium phosphate buffer (pH 7·0) + 0·005 M sodium ethylene diaminetetra-acetate (EDTA) + 0·02% sodium azide for all viruses except the Thai strain of sacbrood virus (TSBV) and Arkansas bee virus (ABV). The best agar for TSBV has the same formulation + 1 to 2% NaCl. A suitable agar for ABV is 0·04 M sodium borate (pH 7·0) + 0·85% NaCl + 0·02% sodium azide.

4. BACTERIA

Bacteria are unicellular microscopic organisms without a nuclear membrane surrounding their genetic material and also without the other nuclear structures and organelles that are in cells of higher organisms. Accordingly, they are known as ''procaryotic'' organisms, whereas fungi and more complex organisms are ''eucaryotic''. Bacteria have cell walls which give them some rigidity and characteristic shapes but the other procaryotic organisms, the mycoplasmas and spiroplasmas, are delimited by a membrane only and, therefore, are more pleomorphic than bacteria. Most bacteria can be cultivated on artificial media and most are beneficial saprophytic organisms. They are ubiquitous and occur in immense numbers and variety, but comparatively few cause disease. There are only four well-known bacteria or bacterial groups that are pathogens of insects, and two of them attack honey bees.

Various strains of *Bacillus thuringiensis* are well-known pathogens that can kill a wide range of the larvae of lepidoptera. The strain that is most pathogenic for the silkworm was identified first, about the beginning of the century. Other strains have shown some promise as agents for controlling wax moths (Chapter 10, VI.A.). Another group of bacilli, of which *Bacillus popilliae* and *Bacillus lentimoribus* are best known, cause ''milky disease'' of the ground-dwelling larvae of certain beetles.

Many other bacteria occur in insects, but most are commensals, or are secondary invaders in diseased individuals, either as saprophytes or as weak pathogens. Sometimes they are uncommon or of dubious nature.

Similarly in honey bees, there are two well-known and widely distributed bacterial diseases, ''American foulbrood'' and ''European foulbrood'', and there are numerous bacteria which are mostly harmless commensals, saprophytes or uncommon pathogens.

I. AMERICAN FOULBROOD

A. Symptoms and Diagnosis

American foulbrood is a disease of larvae which almost always kills them after they have spun their cocoons and stretched out on their backs with their heads towards the cell cappings. These are usually propupae but some pupae die too. They then turn brown, putrefy and give off an objectionable fish-glue-like smell. After about a month they dry down to a hard adherent scale (Fig. 30a) (White, 1920a). The average time before an infected larva shows signs of disease is 12·5 days after hatching, with almost all diseased larvae becoming visibly discoloured between 10 and 15 days after hatching (Park, 1953). The cappings over such larvae quickly become moist and dark-coloured; they then sink inwards and adult bees begin to remove them, first forming small holes and finally leaving the cell fully open. When a matchstick is thrust into the larval remains at the sunken capping stage and then removed, it draws out the brown, semi-fluid remains in a ropy thread (Fig. 30c).

Dry scales fluoresce strongly in ultraviolet light which can help diagnosis with badly preserved material. When a dry scale is placed in 6 drops of milk warmed to about 74°C the milk curdles in about 1 min and then begins to clear at once, all the curd dissolving after 15 min. The effect is caused by stable proteolytic enzymes liberated by *B. larvae* (see below) when it sporulates (Holst and Sturtevant, 1940). It is not caused by any material likely to be tested from colonies, other than by scales of larvae that have died of American foulbrood or by stored pollen, which causes curdling and may appear to cause subsequent clearing (Patel and Gochnauer, 1958). A simpler test is to macerate a little suspect material with 2 drops of milk on a glass slide. Most American foulbrood material produces a firm curd in less than 40 s; European foulbrood material takes at least 1 min 47 s, and healthy larvae take at least 13 min. However, scales may give negative results when they have been in combs which have been fumigated with formaldehyde or paradichlorbenzene, and sometimes for unknown reasons. Dead larvae that have not reached the "ropy" stage do not give a positive reaction (Katznelson and Lochhead, 1947).

B. Cause

American foulbrood is caused by *Bacillus larvae*, a rod-shaped bacterium about 2·5–5 μm by 0·5–0·8 μm. It is motile with peritrichic flagella and is Gram-positive. It forms oval endospores which measure about 1·3 × 0·6 μm (Fig. 31h). These are very resistant to heat, chemical disinfectants and desiccation for at least 35 years (Haseman, 1961). The bacillus appears to be specifically associated with the honey bee and attacks the larvae of workers, queens and drones. Microscopic examination of larval remains shows masses of oval spores with no other organisms. "Giant whips", the coalesced flagella of vegetative rods (Fig. 31i), are readily seen under dark-ground illumination or by phase contrast.

C. Occurrence

American foulbrood occurs in the temperate or sub-tropical regions of all continents, and in New Zealand, Hawaii and some of the West Indies, but not in Africa south of the Sahara (Smith, 1953, 1960). Incidences reported have been about 1% in England and Wales (Ministry of Agriculture, 1969), 3–5% in Switzerland (Tarr, 1937), 7% in Tasmania (Ryan and Cunningham, 1950) and 10% in parts of the U.S.A. (Crane, 1954). The disease has also been reported in *Apis cerana* in India (Singh, 1961).

D. Multiplication of *B. larvae*

Larvae become infected by swallowing spores which contaminate their food. Millions of spores are required to infect a larva older than 2 days, but larvae up to 24 h become infected with about 10 spores or fewer (Woodrow, 1942). Vegetative cells of *B. larvae* are not infective (Tarr, 1937b; Woodrow and Holst, 1942) probably because the food surrounding the young larvae has a bactericidal effect, which Holst (1946) demonstrated on *B. larvae* in artificial culture. The bactericidal effect, which is in part, if not entirely, due to 10 – hydroxydecenoic acid (Blum, Novak and Taber, 1959), decreases when the acid food is neutralized in the larval intestine where the pH is about 6·6.

The spores probably germinate soon after they enter the larval gut. Rods, presumed to be vegetative forms of *B. larvae* were found by Woodrow and Holst (1942) in larval intestines within 24 h of the larvae having their food inoculated with spores. However, they do not multiply in the lumen of the intestine, but eventually penetrate to the haemolymph and then multiply abundantly. Rods were found by Jaeckel (1930) in the haemolymph of 2- or 3-day old larvae, but these were probably severely infected and seem to have been moribund, if not dead. The bacteria may not usually penetrate the gut wall until the larva pupates. For example, in other observations, rods were found after 24 h in the gut of day-old larvae infected with moderate numbers of spores. Few were to be seen there, none were elsewhere when the larvae were between 3 and 6 days old and none could be found anywhere in the larvae when they were between 7 and 9 days old; but the tissues of all larvae between 13 and 14 days old were severely infected (Kitaoka, Yajima and Azuma, 1959).

It seems, therefore, that conditions are optimal in the youngest larvae for germination of spores but soon become unsuitable for vegetative growth. This corresponds with the characters of *B. larvae* grown in culture; spores germinate best in a low oxygen tension, but vegetative growth and sporulation need more aerobic conditions (Section I.F.). The vegetative rods are unable to multiply much in the larval intestine, possibly because it is too anaerobic or bacterial growth quickly makes it so. The rods are motile, however, and many migrate to the gut epithelium, possibly because conditions there are most aerobic (Fig. 31j). Spores seem able to germinate in old larvae, although less readily than in young ones (Bamrick, 1967) and the vegetative rods may not have time to reach the gut epithelium and invade the tissues, before they are evacuated in the faeces along with the gut contents. Larvae of queens are more susceptible to infection than larvae of workers of the same genotype, which in turn are more susceptible than larvae of drones (Rinderer and Rothenbuhler, 1969).

The bacillus proliferates in the tissues of larvae when they become quiescent before pupation. Infected larvae then quickly die and spores form, mostly in propupae 11 days after hatching. About 2500 million spores form in one individual (Sturtevant, 1932) and they are invariably in pure culture. Secondary organisms are unable to grow probably because of the antibiotic released by *B. larvae* as it sporulates (Holst, 1945).

An agglutinin which precipitated cells of *B. larvae* (Gary, Nelson and Munro, 1949) has been found in the haemolymph of both larvae and adults of a colony with American foulbrood. However, it also agglutinated cells of *Bacillus subtilis*. Tests were not made with healthy colonies which may have non-specific agglutinins and other anti-bacterial substances that normally abound in the blood of insects (Wagner, 1961), although laboratory experiments by Gilliam and Jeter (1970) indicated that adult bees injected with spores produced specific agglutinins.

E. Spread of *B. larvae* within the Colony

There is no obvious seasonal outbreak of American foulbrood: disease occurs at any time of the year when brood is present and it has the reputation of invariably killing the colony. However, colonies may show some disease and then recover for an indefinite period (White, 1920), although when more than about 100 cells of dead larvae are seen at any one time the colony is likely to succumb (Woodrow and States, 1943). Spores may be transmitted to larvae by adult bees engaged in cleaning combs, or the larvae that subsequently occupy the same cells may become infected by spores that still remain. However, reinfection of larvae reared immediately in cells that have just previously contained dead larvae is surprisingly infrequent. In one case only 8% became diseased (Woodrow and States, 1943), and in another 19%, mostly in severely diseased colonies. Therefore, spores and remains of dead larvae are removed efficiently by adult bees, and it seems that most infection is transmitted to larvae in other cells. This is probably done by the bees mainly engaged in cell-cleaning but in process of changing their occupation to brood-rearing.

Infected larvae can be detected by adult bees very soon after infection. In some tests, 10–40% were removed by nurse bees before their cells were sealed over, according to the number of spores (1–105) with which they were infected (Woodrow, 1942). In other tests about 50% were removed later, but before they were 11 days old, i.e. before most of them contained spores (Rothenbuhler, 1958; Woodrow and Holst, 1942).

A variety of tissue cells seems to increase in size, and oenocytes multiply in infected larvae before they are invaded by bacteria (Jaeckel, 1930). Possibly the insecticidal factors found in extracts of larvae killed by *B. larvae* (Patel and Gochnauer, 1959), are released by the multiplying bacilli and cause these and other effects, which are detected by adult bees.

Infection may sometimes be eliminated by the action of adult bees, but it seems unlikely that all spores will be removed from a colony once it has had American foulbrood. Spores remain infective for at least 35 years (Haseman, 1961), and some could easily become lodged in unused parts of the nest or in food stores, which may remain untouched for many years.

Diseased colonies that recover seem to do so best during good nectar-flows (Ibragimov, 1958; Reinhardt, 1947). Out of 14 diseased colonies observed by Ryan and Cunningham (1950) in Tasmania, 8 had less disease and 4 had shown no increase after the main nectar-flow. Spores may be so diluted by the incoming nectar that young susceptible larvae have little chance of receiving them in their food. Bees also avoid cells that contain visible remains of dead larvae when storing their nectar or pollen (Woodrow, 1941) and nectar flows stimulate the hygienic behaviour of the older bees of colonies that have been selected for the ability of their nurse bees to detect and remove diseased larvae (Thompson, 1964). Pollen collected by bees protects young larvae from infection to some extent when added to their food (Rinderer *et al.*, 1974). This may well happen when pollen is abundant as it usually is during nectar-flows. Moreover, up to 80% of spores that contaminate nectar are removed by the proventriculus of adult bees (Sturtevant and Revell, 1953), so relatively few become lodged in honey. When 500 million spores/day were fed in 1 litre of syrup for 10 days to colonies allegedly susceptible to disease, followed by 5000 million spores/day after a lapse of 20 days for a further 20 days, only 1 or 2 of several thousand individually identified larvae became visibly diseased and only about 10% of the larvae were ejected (Thompson and Rothenbuhler, 1957). Out of 187 samples of commercial honey in the United States only 15 contained spores and only 1 had more than the 50 million spores/litre found necessary to cause disease by feeding (Sturtevant, 1932).

Contrary to popular belief, therefore, the natural rate of spread of infection of *B. larvae* is low, mainly because most spores are removed from circulation by adult bees and because only the youngest larvae are susceptible.

Some, perhaps many, infected colonies survive with little evidence of disease. However, when the disease kills a few hundred larvae in a colony the infection then usually spreads quickly and the colony dies.

F. Cultivation of *Bacillus larvae*

Bacillus larvae will neither germinate nor sporulate on ordinary bacteriological media. However, an inoculum of very few spores will germinate in semi-solid agar of the following medium when it is inoculated while molten and incubated at 34°C: 1% yeast extract ("Difco"), 1% glucose, 1% starch, 0·136% KH_2PO_4, adjusted to pH 6·6 with KOH and autoclaved at 116°C in closed screw-capped tubes for 20 min. The spores germinate between 5 and 10 mm below the surface in a few days and vegetative growth later extends to the surface. The vegetative cells may then be transferred to agar plates of the same medium but without glucose, and sporulation occurs within a few days (Bailey and Lee, 1962).

Extract of larvae (White, 1907), unheated egg yolk (White, 1920), and egg yolk with yeast, carrot extract and peptone (Sturtevant, 1932) have also usually proved suitable media, although inocula of many millions of spores have been found necessary to start growth on them (Tarr, 1937c). A simplified medium of glucose-peptone, and thiamine plus trace elements has been found satisfactory (Lochhead, 1942) although it was later found necessary to add soluble starch to it, or to extract it with activated charcoal to get reliable growth (Foster, Hardwick and Guirard, 1950).

Virulence or infectivity of *Bacillus larvae* decreases rapidly after it has been cultivated on artificial media (Tarr, 1937b). Some of the cultural characters of *B. larvae* are as follows: colonies are whitish, somewhat transparent and slightly glistening; nitrites are usually produced from nitrates (Hitchcock and Wilson, 1973); purine bases are essential for growth; thiamine replaces some of the essential growth factors of vegetable or yeast extracts; acid but no gas is produced from xylose, glucose, fructose, galactose, salicin and sometimes from lactose and sucrose; acid and curdling form in carrot-milk; and carrot-gelatin is slowly liquified (Breed, Murray and Smith, 1957).

II. EUROPEAN FOULBROOD

A. Symptoms and Diagnosis

European foulbrood is a disease of larvae (Fig. 33) which kills them usually when they are 4 or 5 days old, mostly in early summer when colonies are growing rapidly. There is often a well-defined seasonal outbreak followed immediately by a spontaneous recovery (Morgenthaler, 1944; White, 1920b) (Fig. 8). The sick larvae first become displaced in their cells, just as they do when they are deprived of sufficient numbers of adult bees to feed them adequately. They then soon die, become flaccid, turn brown and decompose, often giving off a foul odour or a sour smell, but sometimes having little or no smell.

Before they decompose, diseased or dead larvae can be dissected easily on a microscope slide by grasping the cuticle at the centre of the body with two pairs of forceps which are then pulled apart. The mid-gut contents are left exposed on the slide still within the transparent gelatinous peritrophic ''membrane'' which is partially, or completely, filled with bacteria in opaque chalk-white clumps. The contents of the mid-guts of healthy larvae, which are less easily dissected, appear golden-brown.

B. Cause

European foulbrood is caused by *Streptococcus pluton* (Bailey, 1957), which is a Gram-positive bacterium, lanceolate in shape and occurring singly, in chains of varying lengths or in clusters (Fig. 31a, b, c,). Its presence was suspected many years ago, and it was then called *Bacillus pluton* (White, 1912; 1920b). Recently, Bailey and Collins (1981) have shown that the guanine + cytosine content of its nucleic acid is about 29%. This quality alone excludes it from the genus *Streptococcus*, as now defined (Buchanan and Gibbons, 1974) and the organism will probably become the type species of a new genus.

C. Occurrence

European foulbrood has long been recognized in Europe and North America and has been diagnosed more recently in bee larvae from Australia, Japan, South Africa, Finland, Norway, Tanzania, Zambia, South America and in *Apis cerana* from India. All strains of *S. pluton* from *Apis mellifera* and one from *Apis cerana* are serologically indistinguishable or are very closely related (Bailey, 1959a, 1967a, 1974, 1977; Bailey and Gibbs, 1962; Tham, 1978). One strain from *Apis cerana* has been found much more distantly related than the

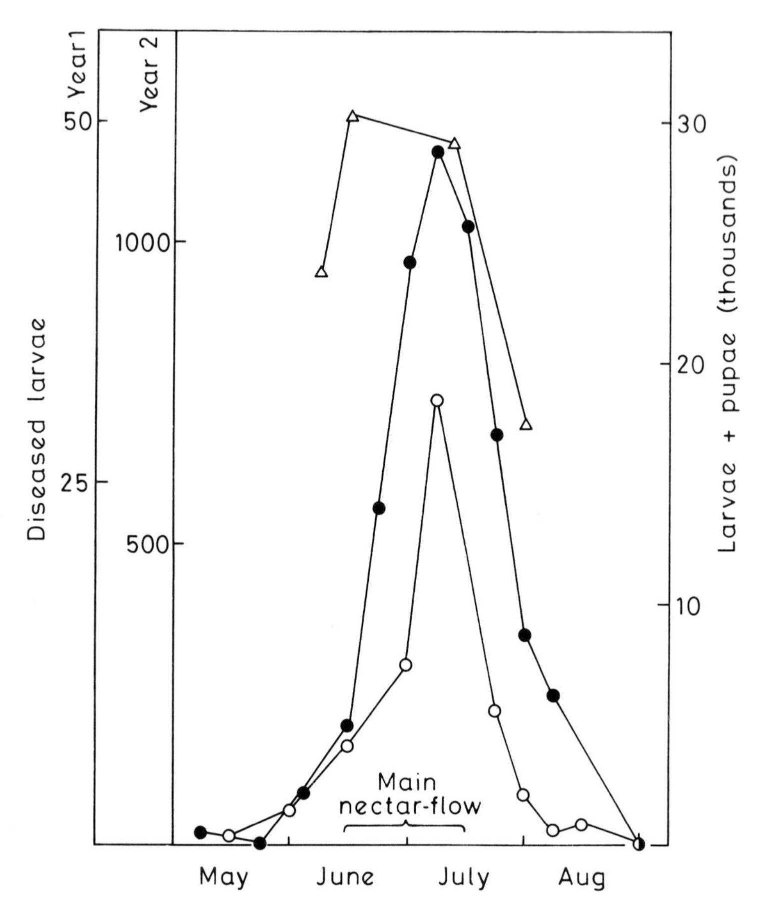

Fig. 8 Natural outbreaks of European foulbrood in undisturbed, untreated, naturally infected colonies. Average numbers of diseased larvae in two colonies during Year 1 (●—●) and the average total number of larvae and pupae in them (△—△); and the average number of diseased larvae in four colonies at the same location in Year 2 (o—o). (From Bailey, 1960).

others to the type strain (Bailey and Collins, 1981). Russian workers have repeatedly referred to *Bacillus pluton* in their work but their cultural methods exclude the possibility that they were dealing with *S. pluton* and there are strong reasons for believing they were working with 'Bacterium eurydice' (Bailey 1959a) (Section IIE). It is likely, however, that the disease they describe is European foulbrood and that *S. pluton* is in Russia. European foulbrood has never been found in New Zealand (Palmer-Jones, 1971).

D. Multiplication and Spread of *S. pluton* within the Colony

The bacteria are swallowed with contaminated food and multiply within the cavity of the mid-gut, which they sometimes fill almost completely. Infected larvae may nevertheless survive and pupate and the bacteria are then discharged with the faeces and deposited on the walls of the brood-combs cells (Bailey, 1959b). The deposits occur mainly at the base and on the capping of the cell and *S. pluton* in them remains infective, possibly for several years. Most bacteria are removed by house-cleaning adult bees, but some find their way to other larvae.

The infected larvae that survive produce pupae of subnormal weight (Fig. 32), because the bacteria have assimilated much of their food (Bailey, 1960a). Larvae are susceptible to infection at any stage of their unsealed life but the older they are the less they are affected, because they lose proportionately less food to the multiplying bacteria. Infected larvae spin feeble cocoons, having poorly developed silk glands, and this encourages the spread of bacteria from their faeces, which otherwise would become sandwiched between layers of cocoon and so removed from circulation.

A balance can exist in an infected colony between the production and dissemination of *S. pluton*, mainly from infected larvae that survive, and its elimination by nurse bees which clean out cells and eject many infected unsealed larvae, often before these are visibly diseased. Infection may persist this way for many years and cause little or no obvious disease. However, the amount of infection fluctuates, and the fluctuation is related to changing availability of hypopharyngeal protein food from nurse bees. Experiments have shown that infected colonies artificially deprived of much of their unsealed brood keep proportionately more infected larvae than usual (Bailey, 1960). This is because larvae which are left receive surplus food when much brood is removed from colonies (Gontarski, 1953; Simpson, 1958). Conversely, more infected larvae than usual are ejected by bees from colonies with European foulbrood when unsealed brood is added, because infected larvae are the first to show signs of starvation when there is a sudden shortage of glandular food.

In endemically infected colonies in spring, before the seasonal outbreak of disease, there is a balance, as described above, between the increase and spread of *S. pluton* and its elimination by the ejection of infected larvae. When inclement weather or other events interrupt nectar-flows, growth of colonies is checked, and *S. pluton* can then accumulate with little or no sign of disease because of a temporary abundance of nurse bees and of the food they provide both for larvae and bacteria. When the main nectar-flow begins, brood-rearing suddenly increases (Simpson, 1959) and severely infected larvae then have insufficient glandular food. When this change is rapid, hastened in all probability by the recruitment of young bees to foraging duties, many infected larvae die faster than bees can detect and eject them, thus creating a typical outbreak. Such events led to the detection of European foulbrood for the first time in Africa (Bailey, 1960) when nectar-flows began unusually late in Tanganyika, after prolonged dry weather had restricted brood-rearing. However, the death and ejection of severely infected larvae prevents further transmission of *S. pluton* in their faeces and this, together with the end of colony growth when the nectar-flow ends, usually soon leads to the disappearance of visible disease. The sequence was well illustrated by the abrupt, spontaneous increase and decrease of the numbers of larvae with

European foulbrood at the same time of year in the same locality in Britain, during both mild and relatively severe outbreaks of disease, which were coincident with the main nectar-flow and the concomitant peak of brood-rearing (Fig. 8).

Small colonies grow relatively more quickly than large ones in the same favourable conditions, which explains the sudden prevalence of disease often reported among "weak colonies". Prolific queens seem to produce resistant colonies, probably because their brood-nests remain larger in proportion to the number of nurse bees than those of less prolific queens, particularly in adverse circumstances, and so their infected larvae are likely to be detected and ejected. Diseased colonies that are near to starving have been seen to lose all signs of disease (Poltev, 1950), because they eject starving larvae and infected ones are the first to show signs of starvation.

Sometimes colonies are destroyed or seriously crippled by disease when sufficient numbers of *S. pluton* accumulate to kill a large proportion of the brood. The death of many larvae makes more glandular food available for the relatively few remaining larvae, enabling the infected ones of these to produce even more bacteria, and the brood-nest soon becomes choked with decomposing remains. However, most infected colonies are not so severely affected, and they usually end good seasons with surplus honey and no apparent disease. In localities with uninterrupted nectar-flows, in which colonies grow unhindered each year, infection usually remains slight, and its effect on most colonies is transient.

E. Secondary Bacteria

The death of infected larvae may be accelerated by secondary bacteria, of which the commonest is "Bacterium eurydice" White. This has its source in the alimentary tract of adult bees and occurs commonly in healthy larvae in small numbers, but is much more numerous in larvae already infected with *S. pluton*. Its incidence is low in winter and early spring, but it becomes abundant in summer. Freshly collected pollen becomes contaminated with it, but the organism dies within a few days in stored pollen. It will multiply in a watery extract of pollen and increases in bees in summer, probably when they are collecting, eating and manipulating pollen (Bailey, 1963b). It takes the form of square-ended rods, single or in chains, or of streptococci, according to its culture medium, (Fig. 31e, f) and has often been confused with *S. pluton*. The cultural characteristics of "B eurydice" have been compared statistically in detail with those of many other bacteria by Jones (1975) who found they most closely resembled those of *Corynebacterium pyogenes*, a pathogen of vertebrates.

A fairly common secondary invader is *Streptococcus faecalis* (*Streptococcus apis* Maassen), which is very like *S. pluton* in appearance (Fig. 31d) and has frequently been confused with it, although the bacteria are serologically and culturally distinct (Bailey and Gibbs, 1962; Gubler, 1954). It does not persist in honey bee colonies by itself, but is brought in by bees from the field where it is very common (Mundt, 1961) and it becomes temporarily established where European foulbrood breaks out. It causes a sour smell, hence the German name, *Sauerbrut*.

Another common secondary organism is *Bacillus alvei* originally believed to cause European foulbrood (Cheshire and Cheyne, 1885), but soon recognized to be a saprophyte living on the dead remains of larvae and not always present (White, 1906). It forms very resistant spores (Fig. 31g) and becomes established in colonies that have had endemic European foulbrood for several years: it causes a characteristic foul odour. *Bacillus laterosporus* also occurs very occasionally in a similar capacity.

Larvae that die decompose rapidly, and many are left by the adult bees to dry to a scale. Only *S. pluton* and spores and *B. alvei* stay alive for long in the scales. Sometimes propupae die after they have voided their gut contents. Few bacteria are left in their remains, but when *B. alvei* is one of them it multiplies quickly and the scale then becomes full of its spores, which are in almost pure culture.

B. alvei has been shown to multiply in propupae killed by sacbrood, after it had been placed in the food of the larvae together with sacbrood virus (Bailey, Fernando and Stanley, 1973). This was analogous to its behaviour in European foulbrood. It failed to multiply when it was fed alone to the larvae. The secondary organisms described above, and others, have at one time or another been considered pathogenic for larval honey bees and to cause European foulbrood, but without experimental evidence.

The natural history of infection of larvae by *S. pluton* and other organisms is illustrated in Fig. 9. For reasons already discussed, larvae infected with *Streptococcus pluton* follow one of four courses of events:

1. They are detected before they are capped and are ejected by nurse bees. *S. pluton* is alone, or is the dominant organism.
2. They die before they are capped and before they are detected by nurse bees. Infection by *S. pluton* is severe and secondary organisms multiply quickly.
3. They are capped and fail to pupate, but usually void most of their intestinal contents. When *Bacillus alvei* is present it multiplies in the remains.
4. They pupate, form normal or undersized adults, and leave infective cells of *S. pluton* in their faecal deposits in the cell.

F. Cultivation of *Streptococcus pluton*

Streptococcus pluton may be isolated from diseased or dead larvae or their cell cappings (section IID), or ideally from dry smears of diseased larval mid-guts (Bailey, 1959b), on a freshly prepared medium composed of: yeast extract ("Difco" or home-made), 1 g; glucose, 1 g; potassium dihydrogen phosphate, 1·35 g; soluble starch, 1 g. This should be diluted to 100 ml with distilled water, the pH adjusted to 6·6 with KOH, and 2 g agar added and autoclaved at 116°C (10 lb/in^2) for 20 min in closed screw-capped bottles (Bailey, 1957).

Watery suspensions of the natural material may be streaked on plates of the special agar, or, preferably, decimal dilutions should be inoculated in the molten agar at 45°C, which is then poured into plates. The plates must then be incubated anaerobically (e.g. in McIntosh and Fildes jars) with about 10% CO_2 at about 34°C. Small white colonies of

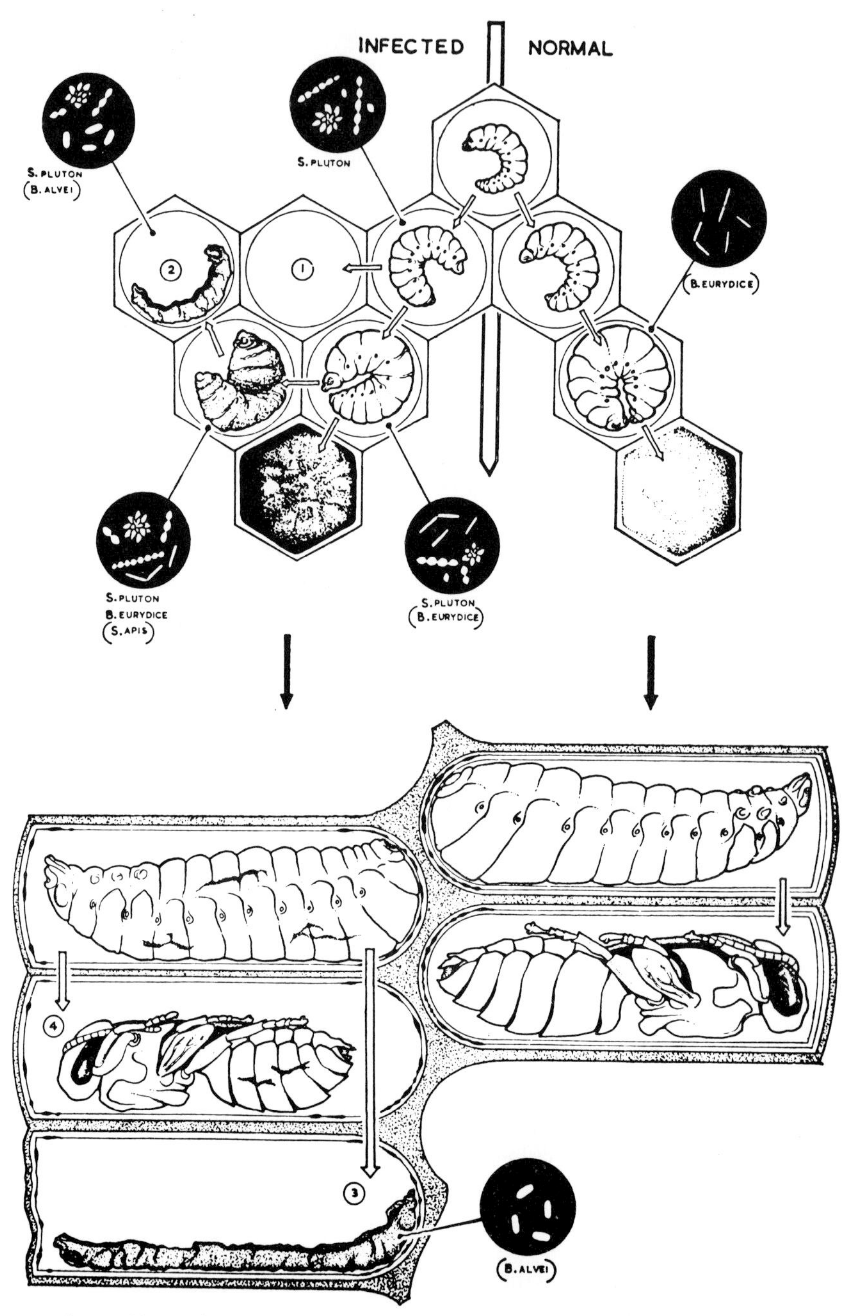

Fig. 9 *The natural history of European foulbrood: pluton.* Names in the diagram are of dominant bacteria; names in parenthesis are of bacteria that do not always occur. See Section II.E. for full explanation.

S. pluton usually appear after about 4 days. *S. pluton* is very pleomorphic in culture, often being in rod-like forms (Bailey and Gibbs, 1962).

Fructose may be used instead of glucose in the medium but, apart from this modification, or very slight modifications in the proportions of the constituents (Glinski, 1972), no other laboratory conditions are known in which *S. pluton* will multiply. Recently Bailey and Collins (1981) have shown that free cysteine is essential, and its incorporation in the medium (0·1 g/100 ml) enables *S. pluton* to multiply with most kinds of yeast extract. Certain peptones, e.g. "Bacto" or "Oxoid", are satisfactory instead of yeast extract, but added cysteine is still either essential or much improves the multiplication of *S. pluton*. Other peptones will not support growth, even with added cysteine.

If the apparatus for anaerobiosis is lacking it is usually easy to cultivate *S. pluton*, when few other organisms are present in the natural material, in deep agar of the special medium. This should be incubated at 34°C, preferably in a closed jar to which some CO_2 is added. When the inoculum is almost free of secondary organisms, particularly *S. faecalis*, growth of *S. pluton* usually occurs below about 8 mm from the surface of the agar.

"B. eurydice" and *B. alvei* grow feebly under the culture conditions necessary for *S. pluton*. *S. faecalis* grows luxuriantly, however, but it can usually be diluted out from natural inocula. Otherwise it produces too much acid, and *S. pluton* cannot grow below a pH of about 5·5.

The presence of *S. faecalis* or *B. alvei* is presumptive evidence of European foulbrood, but their absence is not certain evidence that European foulbrood is absent. Both organisms grow well aerobically on ordinary bacteriological media. *S. faecalis* produces small transparent grey colonies within 24 h. Its growth is much stronger with any one of various carbohydrates in the medium, and the final pH is then about 4·0. *B. alvei* grows very quickly and produces a spreading growth of transparent colonies some of which are motile and move in arcs over the agar surface. The cultures have a characteristic odour and spores form promptly. Further details of the cultural characteristics of these organisms are given in standard works (Gibson and Gordon, 1974; Deibel and Seeley, 1974).

III. OTHER BACTERIAL INFECTIONS

Several bacteria have been isolated from adult honey bees, usually from their intestinal tracts (Kluge, 1963), but very little is known of their natural history.

The intestines of newly emerged bees are free of bacteria; adult bees between 1 and 14 days old, which eat much pollen, have many bacteria in their mid-guts, but later, when their diet is mostly honey, the bacteria almost disappear. However, within the first 2 days of the adult bee's life the pylorus and hind-gut become permanently colonized by bacteria, which grow close to the surface of the epithelium. In the hind-gut the bacteria are mostly short rods, but in the pylorus there is a variety of forms: cocci, short and long rods, and very long filamentous forms. Some may be forms of "Bacterium eurydice". Drones generally have no bacteria in the pylorus; the few that have are also infected with flagellates (Chapter 6, IV.). Queens, too, usually seem to have no bacteria in the pylorus; only 1 out of 16 examined was found to have them, and her pylorus was also infected with flagellates (Lotmar, 1946).

Pseudomonas apiseptica and other bacteria have been found in the haemolymph of bees found dying in or near colonies in various parts of the North America, Europe and Australia. Bees killed by the bacteria sometimes fall apart at the joints of their bodies. The bacteria grow well on ordinary bacteriological media. Bees are not easily infected when fed with the bacteria or when they are in contact with infected individuals, but they can be more readily infected by brief immersion in a watery suspension of the organisms. Wetting the head and thorax seem more effective than wetting the abdomen (Wille, 1961). This suggests that infection occurs via the tracheae, because inhalation is mostly through the first pair of thoracic spiracles (Bailey, 1954). However, no one has infected colonies of bees experimentally (Zeitler and Otte, 1967) and naturally occurring disease seems transitory. *P. apiseptica* has been isolated "in abundance" from soil near infected apiaries (Burnside, 1928), and may really be a soil organism occasionally able to infect bees. Its ready growth on ordinary bacteriological media suggests that it is usually a saprophyte and that it may not be specifically associated with honey bees. The suspicion remains that, in natural circumstances, the various agents of septicaemia cause secondary, though fatal infections, that are perhaps consequent upon a variety of primary pathogens.

Bacillus pulvifaciens (Katznelson, 1950) seems to cause a rare disease, "powdery scale", of larvae. The larvae form a light brown to yellow, very crumbly scale, but most infected larvae are detected and removed by bees before infective spores form (Hitchcock *et al.*, 1979). The bacterium grows and sporulates on ordinary media and could well be a saprophyte that is a fortuitous and ill-adapted pathogen of bees. Many of its cultural characteristics resembled those of *Bacillus larvae* (Gibson and Gordon, 1974).

Spiroplasma spp. are helical motile forms of the class mollicutes, which are procaryotic organisms lacking a cell wall. Several forms of spiroplasma are known, or suspected, causes of certain diseases in plants and insects and one has been isolated from the honey bee (Clark, 1977). It was in the haemolymph of a few moribund worker bees in Washington D.C., and was cultivated on standard media used for *Mycoplasma* spp., and in a mosquito tissue culture medium. Apparently, workers and queens can be readily infected by feeding or injecting them with the cultured organisms. It was found only in early summer and is believed to come from certain spring-flowering plants (Davis, 1978), but it is not clear whether it is primarily a plant pathogen or an insect pathogen that is spread via the nectaries of plants by nectar-feeders.

Wille (1967) identified "rickettsiae", many of which were probably particles of filamentous virus (Chapter 3, I.D.) according to Clark (1977, 1978), in about 60% of sick or dead colonies in Switzerland. However, Wille observed rickettsia-like particles in company with no other pathogen, including *Nosema apis*, in about 25% of several thousand bees investigated. This suggests that many of the particles were not of the filamentous virus, as this is closely associated with *N. apis* (Bailey, 1981), so they may well have been organisms resembling rickettsiae. True rickettsiae are minute bacteria-like organisms that are obligate, intracellular parasites of the alimentary tract of blood-sucking arthropods.

5. FUNGI

The fungi, or Eumycetes, include the moulds and yeasts and most are saprophytic on decomposing organic matter. However, the first disease of any kind that was correctly diagnosed as being caused by an infectious micro-organism was a fungal disease of an insect. Agostino Bassi showed in 1834 that "calcino" or "muscardine" of the larvae of silkworms was caused by a fungus, known today as *Beauvaria bassiana* (Steinhaus, 1956). This early knowledge was overshadowed by the strong belief that prevailed about spontaneous generation, but after Pasteur (Chapter 1) it soon became clear that many insects were prey to a variety of fungi. However, as with bacteria, many fungi that multiply in insects are saprophytes. These may be secondary invaders of individuals that have been killed or diseased by primary pathogens, which may also be fungi sometimes obscured by the secondary organisms.

I. CHALK-BROOD

A. Symptoms and Diagnosis

Larvae die of chalk-brood after their cells have been capped. They are at first somewhat fluffy and swollen, taking on the hexagonal shape of the cell, but later they shrink and become hard. By this stage the cappings have frequently been removed by the bees. Some of the dead larvae remain chalky-white but others become dark blue-grey or almost black (Fig. 33d). Drone larvae are frequently more affected than worker larvae, but it is not unusual to find combs with only the worker larvae affected.

Young infected larvae do not usually die or show signs of disease, although many from diseased colonies can be shown to be infected by allowing them to die in the laboratory when their bodies become overgrown with mycelium (Maurizio, 1934). Infected larvae usually die within the first 2 days after they have been sealed in their cells; otherwise they die as propupae. Many of the cells may remain sealed in severely diseased colonies, and the loose, hard, larval remains then rattle when the comb is shaken.

Pollen which is mouldy with *Bettsia alvei* (Section I.B.) may be mistaken for chalk-brood, but this usually occurs too early in spring to be chalk-brood, and the lumps break up rather easily into fragments representing the original pollen loads.

B. Cause

Originally known as *Pericystis apis* but now reclassified and renamed *Ascophaera apis* (Spiltoir, 1955; Spiltoir and Olive, 1955), this fungus infects larvae only and causes chalk-brood. *Ascosphaera apis* is heterothallic, i.e. spores form only where two different strains of mycelium touch each other. Spores form in spherical aggregates within dark brown-green spore cysts ("fruiting bodies") which are about 60 μm in diameter (Fig. 10); they are very resistant and remain infective for at least 15 years (Toumanoff, 1951).

Similar fungi have been found with fruiting bodies, in a single dead pupa of a leaf-cutting bee (Melville, 1944), and in the tube of a mason bee (Clout, 1956). Such infections may not be uncommon, but the fungi of the solitary bees are not necessarily the same as

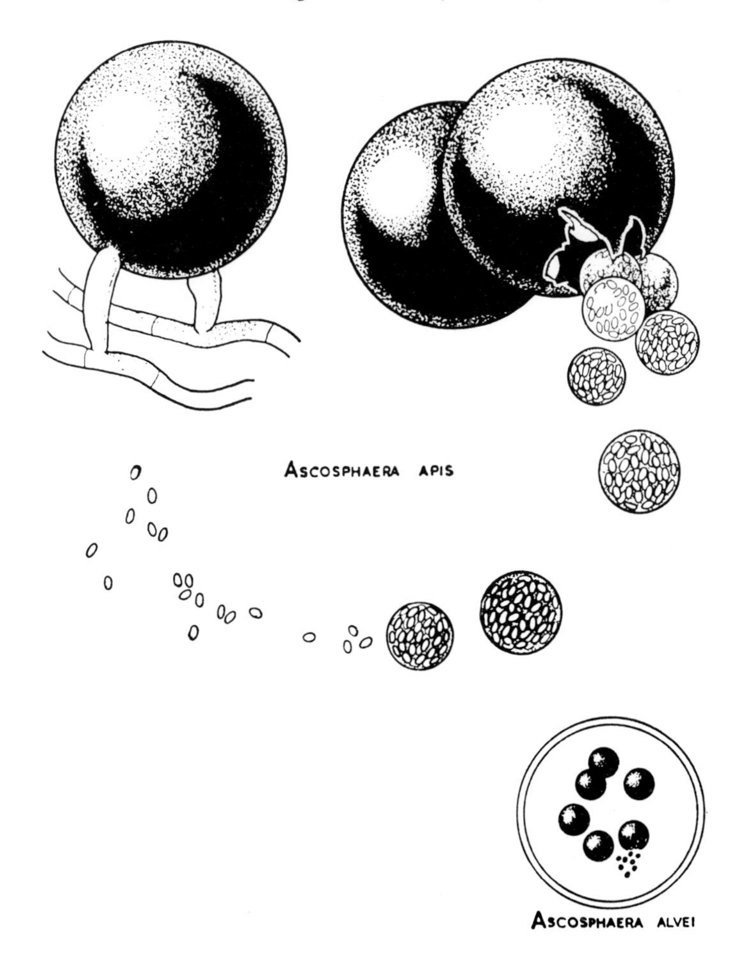

Fig. 10 *Ascosphaera apis*. (After Dade, 1949).

that which attacks honey bees (Skou, 1972, 1979). There may well be several varieties of *A. apis* each specific to a particular species of bee. The *A. apis* found in the mason bee seemed to differ from the honey bee variety as it apparently started its growth on the stored pollen and then moved into the larva.

There is another variety of *A. apis* known to occur in honey bee colonies and with fruiting bodies about 130 μm across (Maurizio, 1935). It was found particularly in larvae killed by European foulbrood, which suggests it may be a saprophyte that acts as a secondary invader of moribund larvae. Its spore cysts are smooth, whereas those of *A. apis* have scattered papillae, visible by scanning electron microscopy (Gochnauer and Hughes, 1976). Skou (1972) has named it *Ascosphaera major* and refers to it as a cause of chalk-brood equally with *A. apis*, although he isolated it from the leaf-cutter bee, *Megachile centuncularis*. Moreover, he observed that it can multiply on the faecal matter of the bees, which suggests it is primarily a saprophyte.

Pollen mould, *Ascosphaera alvei* (Betts, 1912), renamed *Bettsia alvei* by Skou (1972), is a fungus rather similar to *Ascosphaera apis* but its spore cysts are only about 30 μm in diameter and its spherical spores are not aggregated into spore balls. It will grow only at about 18°C and is confined to stored pollen in bee combs.

C. Multiplication

Larvae ingest spores of *A. apis* with their food. The spores germinate and the mycelium begins to grow in the lumen of the gut, particularly at the hind end (Maurizio, 1934). The mycelium then penetrates the gut wall and eventually breaks out of the hind end of the larva's body, often leaving the head unaffected. When they occur, fruiting bodies form on the outside of the dead larvae.

The nature of the infection is strikingly analogous to that by *Bacillus larvae* (Chapter 4, ID). Vegetative growth of *A. apis*, like that of *B. larvae* is poor in the near anaerobic environment of the gut of growing larvae, and *A. apis* depends on the death of its host to form its infective spores. However, *A. apis* grows best in slightly chilled larvae as its optimal temperature for growth and formation of fruiting bodies is about 30°C (Maurizio, 1934). Experiments have shown that brood is most susceptible when chilled immediately after it has been capped (Bailey, 1967C) (Fig. 11). The chilling need be only a slight reduction of temperature, from the normal 35°C to about 30°C, for a few hours; and it can easily occur, even in warm climates, in colonies that temporarily have insufficient adult bees to incubate their brood adequately. Larvae are most likely to be chilled in early summer when colonies are growing, and drone larvae often suffer most as they are generally on the periphery of brood nests. The smallest colonies are at the greatest risk of becoming chilled because they have the lowest capacity for heat and relatively large surface areas from which to lose it. *A. apis* does not multiply in adult bees.

D. Spread

Spread of chalk-brood within a colony is usually limited in nature, perhaps because few larvae usually become infected with both strains of mycelium, and so most do not form infective spores (Maurizio, 1934). However, combs have been found with all the dead larvae infected with the mycelium of one strain (Maurizio, 1934). This was probably derived from a single larva that had been infected with one spore, as it is highly unlikely that each larva was infected with spores all of one or the other strain, and implies that larvae can become infected with mycelium as well as with spores. The main factor that

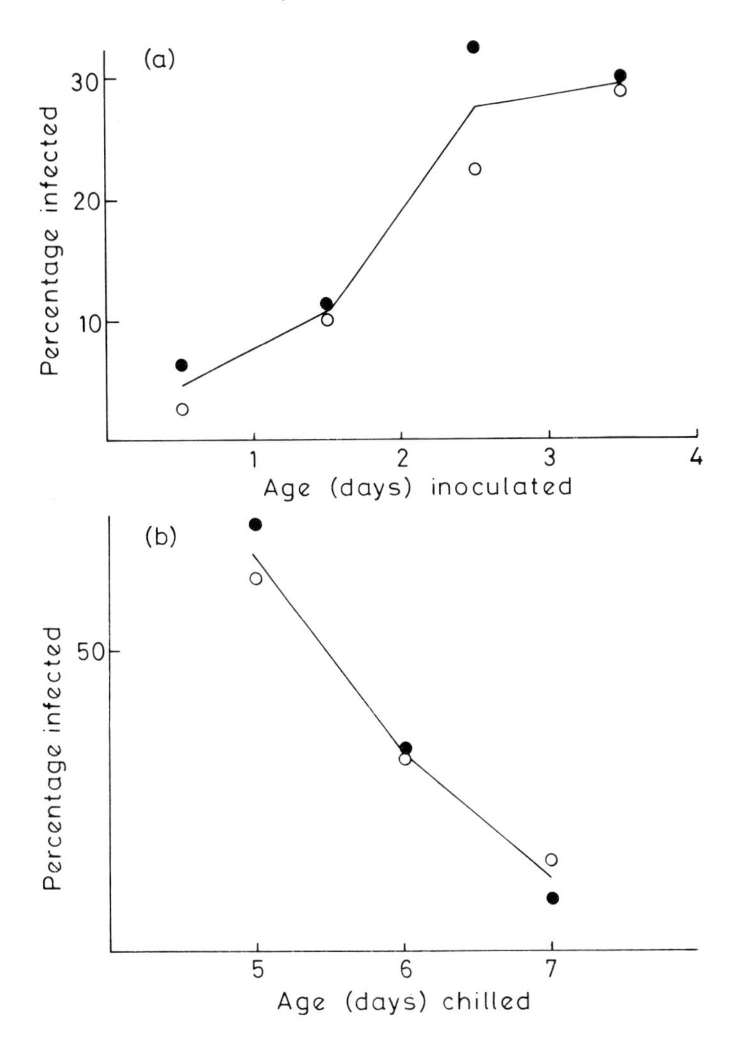

Fig. 11 (a) Percentages of groups of 50 larvae that developed chalk-brood after each larva was inoculated with 10^4 spores at the age indicated and cooled temporarily to about 25°C when 5 days old. (b) Percentages of groups, of 130 or more larvae, that developed chalk-brood after each larva was inoculated with 10^4 spores when 3–4 days old and cooled temporarily to about 25°C at the age indicated; ● Colony 1; o Colony 2. (After Bailey, 1967c).

checks infection in nature is probably the temperature of the brood-nest (Section I.C.; Chapter 10, III.A.).

It is conceivable that other species of bee play a part in spreading infection by *A. apis*. However, even if they can act as hosts for *A. apis* of honey bees, infection will be more likely to pass from honey bee to solitary bee than vice versa, because solitary bees do not touch their own larvae at the age when they would be infective and so could not pass infection to adult honey bees by contact in the field.

E. Occurrence

Chalk-brood occurs widely in Europe, including the British Isles, in Scandinavia and Russia (Betts, 1932) and in New Zealand (Seal, 1957). It was believed to be absent from North America until about 1970, since when it has been found widespread in the mid-western and western United States and in Canada (Gochnauer and Hughes, 1976). The reason for this sudden appearance of chalk-brood in North America is unknown. *Ascosphaera apis* is the cause; *Ascosphaera major* has not been recorded in America.

F. Cultivation of *Ascosphaera apis*

Ascosphaera apis can be cultivated on potato-dextrose agar + 0·4% yeast extract, or on malt agar (0·5 to 2% malt). Strong vegetative growth with aerial hyphae and abundant fruiting occurs on the potato medium, but the malt agar has been found better for microscopic work as aerial hyphae are then virtually absent. Others have found that *A. apis* needs complex nitrogen sources and grows best in media with 0·1% asparagine and 0·5% yeast extract with the pH near neutral but below 7·2. The optimum temperature for growth is about 30°C.

Very small inocula of spores do not germinate well on the media mentioned above but will do so readily in a semi-solid agar of the yeast – glucose – phosphate medium described for the cultivation of *Streptococcus pluton* (Chapter 4, II.F.) and incubated at 35°C (Bailey, 1967c). The mycelium then quickly grows to the surface where spore cysts form, when at least two spores of opposite sign have germinated. Mycelium does not grow below about 15 mm in the medium unless this is incubated at temperatures below 35°C, which enables oxygen to penetrate deeper. This seems analogous to events in the nature, the mycelium being unable to grow vigorously enough in the almost anaerobic gut contents to penetrate the tissues, unless the larvae become chilled.

II. STONE-BROOD

A. Symptoms

Larvae with stone-brood may be sealed or unsealed. At first they are white and fluffy and later turn a pale brownish or greenish-yellow and become very hard. Most die after they have been capped in their cells prior to pupation.

B. Cause

Stone-brood is caused by *Aspergillus flavus* or, less often, by *Aspergillus fumigatus*. Mature growth of *A. flavus* has a yellow-green appearance and that of *A. fumigatus* appears grey-green. Both look similar under the microscope (Fig. 12) and both fungi are common, occurring in soil and cereal products. They infect and kill other insects and sometimes cause respiratory diseases in animals, particularly man and birds.

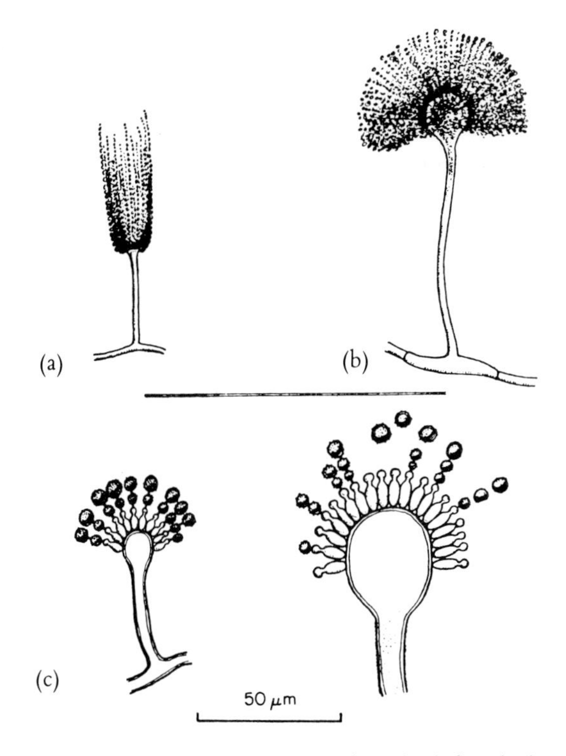

Fig. 12 (a) *Aspergillus flavus*; (b) *Aspergillus fumigatus*; (c) *A. flavus*; detail of conidiophore. (After Dade, 1949).

C. Multiplication and Spread

Spores can germinate on the cuticle of larvae, and the mycelium will then sometimes penetrate the sub-cuticular tissue and produce local aerial hyphae and conidiophores; but it seems that infection is usually via the gut (Burnside, 1930). The internal tissues are quickly overgrown with mycelium which eventually breaks through the cuticle near the anterior end of the body and then grows closely over the cuticle to form a false skin within 2 or 3 days. Conidiophores begin to form at the same time, wherever the mycelium is exposed to the air.

The fungi can multiply in adult bees and undergo a development in them similar to that which occurs in larvae (Burnside, 1930). However, the amount of mycelial growth in the gut of infected individuals, which become flightless and sluggish and crawl away from the cluster of bee colonies, seems inadequate to account for these early symptoms. These are probably caused by toxins liberated by the mycelium. An ether-soluble toxin occurs in cultures of *Aspergillus flavus*, particularly in cultures that have just reached maturity, which is indicated by the deep yellow-green colour of most of the conidia. The extracted toxin is unstable, losing most of its potency within 15 days.

Stone-brood is usually transient, although very large inocula can kill colonies (Burnside, 1930). The death of a few naturally infected colonies has been observed (Dreher, 1953). In these the mycelia had grown into the brood-comb cell walls, probably making it impossible for the adult bees to clean and disinfect cells.

Passage of *Aspergillus flavus* in an insect host raises its virulence (Madelin, 1960), so once established in a colony more virulent strains of the fungus might become selected.

D. Occurrence

Stone-brood is well known in Europe and North America. Infection of larvae by *A. flavus* has been reported from Venezuela (Steiskal, 1958). It is rare in Britain, however, in spite of the common occurrence of the fungi and of the damp climate, which is usually believed to encourage fungal infections. The fungi probably do not usually grow in bees and may do so only when bees are enfeebled by other factors.

III. OTHER FUNGI AND YEASTS

A micro-organism of a primitive type appearing to be transitional between yeasts and fungi has been found in the melanized (blackened) patches of epithelial tissue that surround nurse- and egg-cells of ovaries and the poison sac and rectum in queens: a disease originally called "eischwarzsucht" and renamed "melanosis" (Fyg, 1934). The organism is easy to culture *in vitro*, and adult workers and drones have been infected successfully with cultures injected into the thorax (Fyg, 1936). Bees that ingested the fungus did not become infected. Infected queens soon stop laying and are superseded by a new queen. One unmated queen was found by Fyg to have infected ovaries. The ease with which the organism is cultivated and its power to infect only after injection, suggest that it is a saprophyte which occasionally invades the bodies of bees via wounds.

Similar fungi and symptoms have been described in honey bees and in a solitary bee, *Andrea fulva*, by Orösï-Pal (1936, 1938b). However, Skou and Holm (1980) who investigated queens that were failing to lay many eggs, found many with melanosis but failed to associate the disease with any pathogen. They isolated many different micro-organisms, mostly yeasts, some resembling those described by Fyg, from the reproductive organs of the queens, and found the yeast *Saccharomycopsis lipolytica* the most common; but it was not pathogenic in experiments.

Other yeasts have been isolated on malt-agar from the intestines of sick bees and many bees died in colonies that fed on the yeasts supplied in syrup. Control colonies were unaffected. The pathogenic yeasts seemed to be *Torulopsis* spp. closely similar to *Torulopsis candida* (Giordani, 1952). Many other fungi and yeasts that occur naturally in the gut of adult bees have been listed by Vecchi (1959) and by Skou and Holm (1980), but none seems to be pathogenic.

6. PROTOZOA

Protozoa are seemingly lowly forms of animal life, occurring as microscopic unicellular or acellular organisms. Nevertheless, they comprise a great diversity of types of which many are parasitic or symbiotic, and many of these infect insects. Well-known symbiotes include those that infect termites and that ferment cellulose, enabling these insects to subsist on a diet of wood. Most of the known parasitic protozoa in insects are the spore-forming Microsporidia of which many hundreds of species have been identified in insects of all types (Bulla and Cheng, 1976). One of the best known of these is *Nosema bombycis* which causes "pebrine" in the silkworm, *Bombyx mori* (Chapter 1). *Nosema apis* of the honey bee is another well-known species.

I. *NOSEMA APIS*

Nosema apis develops exclusively within the cells of the epithelium of the mid-gut of adult bees (Fig. 34a). Many research workers have looked for its spores in tissues other than the mid-gut, but have usually found nothing. Microsporidian-like spores have been seen in other tissues, such as ovaries, fat-body and even hypopharyngeal glands of adult bees (Steche, 1960), and have been identified as those of *N. apis*, but only by their size which is an unreliable guide (Kramer, 1960a). Masses of particles that look like spores of *N. apis* and occur occasionally in honey have proved to be starch grains from maize or pollen (Sturtevant, 1919); and some fungal spores, e.g. of the common mushroom, are very like those of *N. apis*. Microsporidian-like spores have also been found in the blood, rectal glands and large flight muscles in the thorax as well as egg and nurse cells and other ovarian tissues and the mid-gut epithelium (Örösi-Pal, 1938b). These spores were seen in histological sections and were very sparse; no infection experiments were made to establish their identity. There may be microsporidia of other insects that sometimes infect honey bees. The species found attacking brood and killing pupae, but not affecting adult bees, in South Africa (Buys, 1972) could well be of this kind. However, *N. apis* is the common type in honey bees.

Claims have been made that other insects can be infected by feeding spores of *N. apis* to them. Fantham and Porter (1913) believed they infected many different species of bumble-bees, wasps, moths and flies this way; and Showers *et al.* (1967) similarly claim to have infected colonies of *Bombus fervidus*. These results require confirmation. It is necessary to ensure that other microsporidia are not already present in the test insects, or that spores seen in the insects are not merely some of the inoculum but have multiplied. Kramer (1964), in well-controlled tests, showed that spores of *N. apis* that were infective for honey bees did not infect muscoid flies, including species that Fantham and Porter (1913) claim to have infected.

A. Symptoms and Diagnosis

Infected bees show no outward signs of disease. Even the mid-gut shows little evidence of damage when infected. The nuclei of infected cells seem normal, although finer details such as the striated border of infected ventricular cells and the formation of peritrophic membranes seem abnormal in severely infected bees (Hertig, 1923), and cytoplasmic granules of calcium phosphate, which are numerous in healthy ventricular cells, disappear in infected cells. However, infected bees live only half as long as non-infected individuals

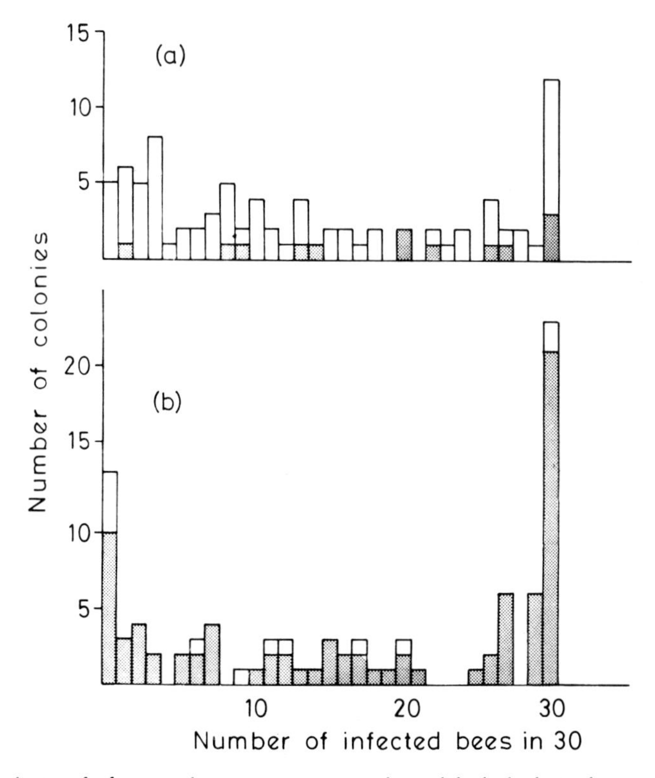

Fig. 13 Distribution of infection with *Nosema apis* among (a) live and (b) dead colonies during March-April, 1963 at Rothamsted; and the distribution (stippled columns) of visible signs of dysentery among them. (From Bailey, 1967e).

in colonies in spring or summer (Maurizio, 1946); the lives of infected caged bees are shortened between 10 and 40% (Beutler and Opfinger, 1949; Bailey, 1958); and infected bees do not fully develop their hypopharyngeal glands (Lotmar, 1936, Wang and Moeller, 1969), which probably explains why about 15% of the eggs in severely infected colonies fail to produce mature larvae in early summer, compared with about 1% in healthy colonies (Hassanein, 1951). Furthermore, infected bees in winter have only about 6 mg of nitrogen in their fat-bodies, whereas healthy bees have been 14 and 23 mg (Lotmar, 1939) and have more amino acids in their haemolymph than infected bees (Wang and Moeller, 1970b). The dry weight of caged infected bees, minus their alimentary canals, decreases more rapidly than that of caged uninfected bees, while their rectal contents gain weight more rapidly, so the bees become dysenteric earlier than uninfected individuals. The dysentery is mainly due to the accumulation of water (Chapter 9, I), since the total water content of infected bees is higher than usual (Lotmar, 1951).

Although *Nosema apis* appears likely to aggravate dysentery, there is no evidence that it is a prime cause of dysentery in nature. A survey of over 100 naturally infected colonies, during a winter when dysentery was prevalent, showed that although it was clearly associated with the death of many of the colonies dysentery was not caused primarily by *N. apis* because it occurred whether or not colonies were severely infected. Moreover, *N. apis* was not the prime cause of death of colonies because it was about equally distributed among living and dead ones (Fig. 13) (Bailey, 1967e). This distribution of *N. apis* was atypical because of the unusual amount of dysentery: characteristically, after most winters, comparatively few colonies become severely infected (Fig. 14).

Infected bees begin duties normally undertaken by older bees, i.e. young infected bees soon cease to rear brood and to attend the queen and they turn to guard duties and foraging (Wang and Moeller, 1970a). This is the same effect on adults as that caused by sacbrood virus (Chapter 3, I.C.4.). The effect of infection on the length of life of caged bees can be offset very largely by providing them with pollen additional to sucrose (Hirschfelder, 1964). Infected queens cease egg-laying and die within a few weeks of becoming infected (Fyg, 1948, L'Arrivée, 1965).

When colonies are artificially infected in spring or summer they recover in a few weeks, but when they are similarly infected in autumn they die in winter (Jamieson, 1955; Morgenthaler, 1941; White, 1919), because there is insufficient time for infection to decrease spontaneously (Section IC) before the bees cluster for the winter. Death of colonies, or serious damage to them, is rarely caused by natural infection, but significant negative correlations have been reported between honey yields and the degree of infection (Hammer and Karmo, 1947; Harder and Kundert, 1951; and Poltev, 1953). In winter, infected colonies lose about 1500 more bees than normal colonies of the same size, regardless of their absolute size, i.e. the smaller the colony the relatively greater the loss of bees from infection (Jeffree, 1955; Jeffree and Allen, 1956). This suggests that small colonies are usually more severely infected than large ones and it might be supposed they are small because of the infection. However, undetermined factors additional to *N. apis*, but correlated with its incidence because they are correlated with dysentery, are much more important as inhibitors of the growth of colonies in spring (Bailey, 1967e).

Many of the pathological effects mentioned above may be due, in part, to one or more of the three viruses that are associated with *N. apis* (Chapter 3, I.D.).

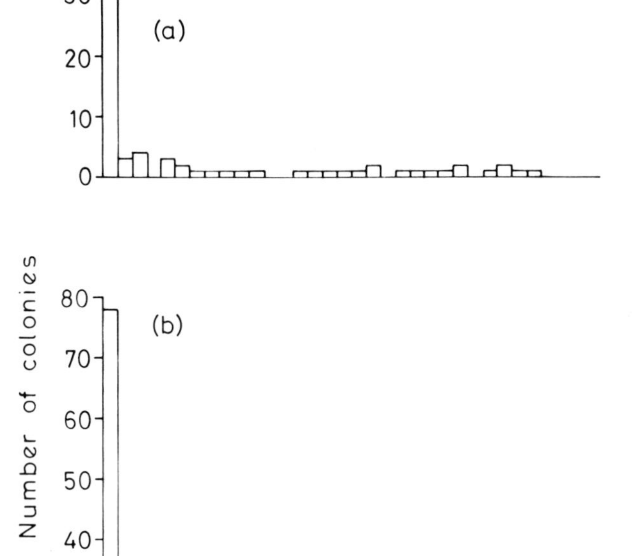

Fig. 14 Distribution of infection with *Nosema apis* among live, seemingly healthy colonies after 2 average winters (a) 1964 and (b) 1965. (From Bailey, 1967e).

Since there are no clear symptoms, the diagnosis of infection depends upon microscopic examination, for the presence of characteristic spores, either of extracts in water of bees, or of their faecal matter. This can be collected on horizontal glass plates mounted near the hive entrance (Wilson and Ellis, 1966); and it can be collected from queens without harm by holding them over glass slides (L'Arrivée and Hrystak, 1964).

B. Multiplication

The spores are ingested by the bee and are passed quickly into the mid-gut by the proventriculus. As soon as they enter the mid-gut they each extrude their hollow polar filament and inject the germ through it into an epithelial cell (Kramer, 1960b, Morgenthaler, 1963), (Figs 15, 34d).

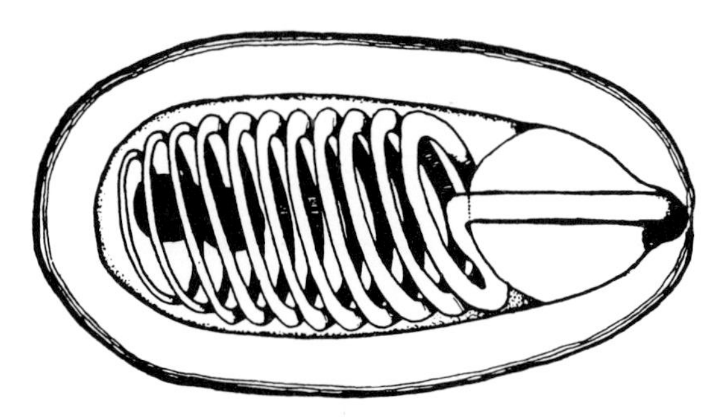

Fig. 15 Internal anatomy of a spore of *Nosema apis*; diagram based on electron micrographs of ultrathin sections, showing the polar filament coiled round the twin nuclei of the sporoplasm. (After Huger, 1960).

The parasite develops and multiplies in the cytoplasm of the host cell, and in bees kept at 30°C the spores form after about 5 days. These are cast into the gut and pass to the rectum, often still inside the host cell (Fig. 34a) when, as in normal bees, this is sloughed off. All the cells of the mid-gut are eventually parasitized possibly by reinfection from newly-produced spores that have been cast off into the gut cavity, or by invasion of vegetative forms from adjacent cells as described for *Nosema bombycis* in the silkworm (Isihara, 1969). About 30–50 million spores are in the gut of a bee when infection is fully developed.

N. apis does not infect honey bee larvae (Hassanein, 1951). Newly-emerged bees are always free of infection, but they are as susceptible as older bees (Bailey, 1955a).

C. Spread

Spores were once believed to be disseminated by the wind, a variety of insects, and the flowers and water at drinking places visited by bees (Fantham and Porter, 1912, 1913, 1914; White, 1919). However, early work showed that infection declined spontaneously during the summer in North America and Europe after a spring peak (Bullamore and Malden, 1912; White, 1919; Morgenthaler, 1939). Suggestions were made that, although spores were being spread, high summer temperatures inhibited their multiplication in bees (Schulz-Langner, 1958). However, temperatures have to be raised to more than 35° for many days to inhibit multiplication of the parasite in bees in the laboratory (Lotmar, 1943) and the temperature in the centre of the brood-nest rarely exceeds this (Ribbands, 1953) even on hot days. Higher temperatures outside the brood-nest could affect only a fraction of the adult bees for brief periods and cannot account for the spontaneous decline of infection in cool climates, such as in Britain. Here, experiments showed that the parasite multiplied as much as ever in marked infected individual bees that were placed in endemically infected colonies, and then sampled at intervals during the period when the percentage of infected bees was declining rapidly and spontaneously (Bailey, 1959). Therefore, the only explanation for the spontaneous decline of infection in summer is that infection does not spread from infected bees to the next generations of individuals.

Spores are spread in the faecal matter of adult bees and are ingested by young individuals when they clean contaminated combs. Bees are more likely to defaecate within the cluster during late winter, after their long confinement, than at other times, because the weight of their rectal contents then increases rapidly (Nitschmann, 1957). The sharp increase in weight is attributed to a heightened metabolism at the start of brood-rearing whilst bees are still mainly confined to the cluster.

When bees are able to fly freely in summer, and defaecate away from the colony the combs become cleaner and the chances of a bee contacting spores decrease. The infected bees die without transmitting their infection, which accordingly decreases. For the same reason spores of *N. apis* rarely occur in the honey or pollen collected by the bees from natural sources.

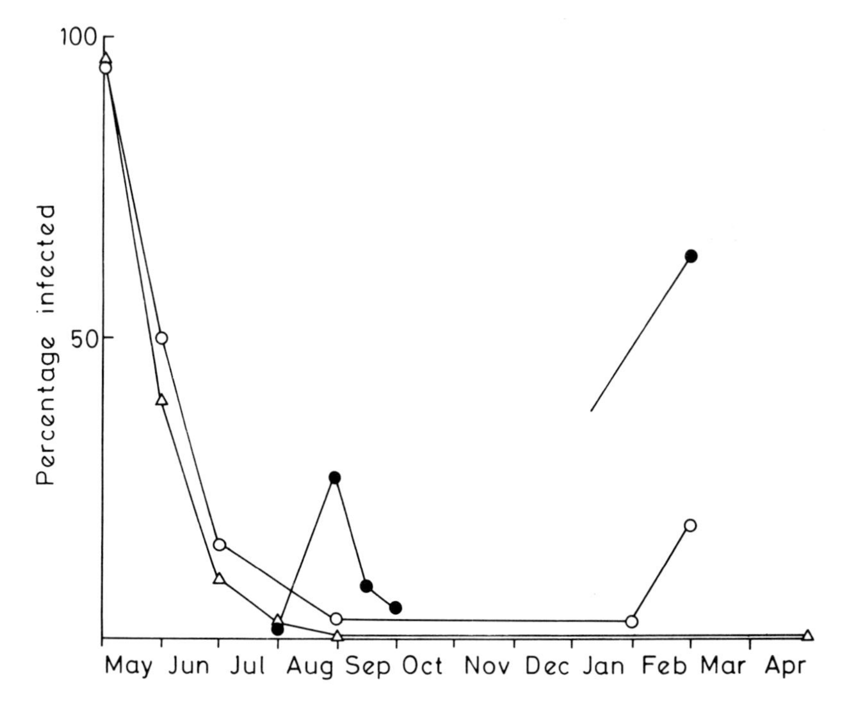

Fig. 16 Average percentage of bees infected with *Nosema apis* in: 3 naturally infected untreated undisturbed colonies (o—o); 1 naturally infected colony transferred entirely to uncontaminated comb in June (△—△); and 2 healthy colonies into which several combs, removed from severely infected colonies in spring, were placed in July (●—●). Each point represents the average number of infected individuals in samples of 100 bees from each colony. (Partly from Bailey, 1955b).

Queens do not clean combs so they rarely become infected under natural conditions. Drones become infected, but significantly fewer than workers although they are equally susceptible: they probably acquire infection by chance when they are fed by workers also engaged in cleaning combs (Bailey, 1972).

Colonies transferred carefully to non-contaminated combs in early summer lose their infection at about the same rate as untreated control colonies (Bailey, 1955b, 1955c), but by the following spring the treated colonies have less, often no infection, whereas the usual increase occurs in untreated colonies (Fig. 16 and Table VIII). Furthermore, combs taken from infected colonies in spring and placed in uninfected colonies in late summer, cause a resurgence of infection (Fig. 16). This dies away at about the same rate as in early summer, but has insufficient time to decrease as much as usual and so reappears in spring even more severely. These observations indicate that some infection persists on combs throughout the summer sufficiently to infect the winter cluster, and tests have shown that spores can survive dormancy in faecal deposits for at least a year (Bailey, 1962).

Infection can remain endemic within a colony, but it is evident that recontamination of the comb by bees virtually ceases in summer. However, defaection by bees can occur in the colony in summer when bees are confined by unusually inclement weather (Borchert, 1948), and there is a statistically significant positive correlation between cool, dull, rainy summers and the infection in colonies the next spring (Lotmar, 1943); although this may be partially because colonies grow slowly in rainy years and clean their combs less thoroughly than usual. Experiments showed that infection in spring was significantly less in colonies in which contaminated combs were put in the centre of their brood-nests during the previous summer, than in colonies in which contaminated combs were put on the periphery of the brood nests (Bailey, 1955b). This is because combs in the centre of the brood-nests are used more, and are therefore cleaned more, than those as the periphery. Colonies do not grow when they lose their queens and they stay more severely infected than usual (Table VIII, Chapter 10, IV.A.3.).

D. Occurrence

Nosema apis occurs in *Apis mellifera* throughout the world. Considerable differences have been reported between its incidences in different countries but the amounts found probably depend greatly on the scale and timing of the investigations. The true incidences are probably considerably greater than the values that have been reported. These range from less than 2% of colonies in Italy (Giavarini, 1956) to more than 60% in the Black Forest regions of Germany (Kaeser, 1954). However, most colonies are usually only slightly infected (Fig. 14). A survey in the U.S.A. showed that most queens in small colonies ("package bees") sent from the south to the northern states in spring were infected (Farrar, 1947); although later, Furgala (1962) found very few, and Jay (1966) found between 4 and 20% only of queens were infected in similar packages, possibly because dealers had begun to take measures against infection (Chapter 10, IV.A.).

II. *MALPIGHAMOEBA MELLIFICAE*

This organism, also known as *Vahlkampfia mellificae* (Steinhaus, 1949) of the order Sarcodina, infects the lumen of the Malpighian tubules of adult bees (Prell, 1926), where it develops first as an amoeba-like individual which ultimately encysts (Fig. 34b, c).

A. Symptoms and Diagnosis

The epithelium of infected Malpighian tubules may atrophy, but no other effects of infection have been found. Diagnosis depends upon the detection of cysts by microscopy, as for spores of *Nosema apis* (Section IA).

The effect of infection on colonies is uncertain; it is probably harmful but there are no known symptoms. The colonies used for the experiments described in Fig. 17 appeared normal at the height of infection. There have been reports of serious and even fatal infections, although the circumstances are not clear (Jordan, 1937).

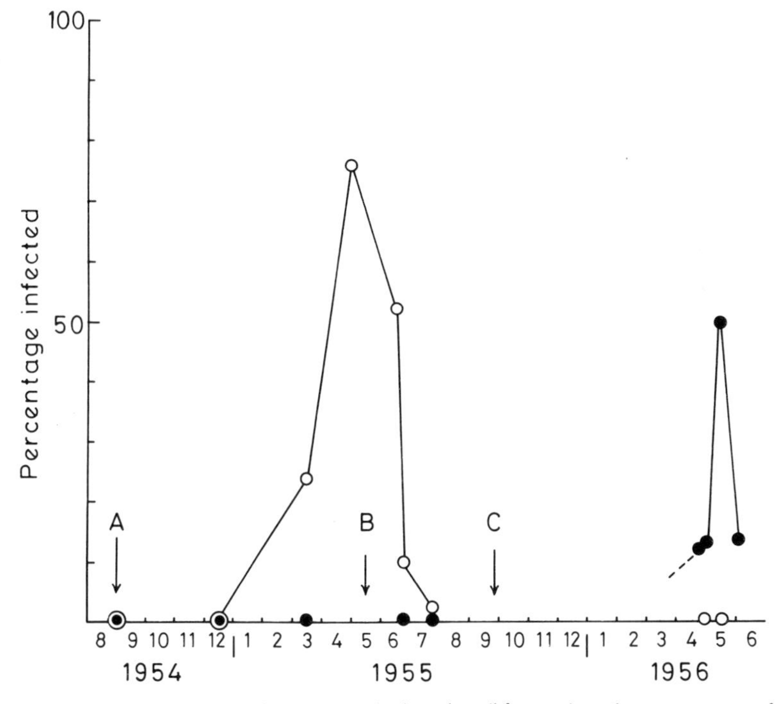

Fig. 17 Average percentage of bees infected with *Malpighamoeba mellificae* in 2 bee colonies. A. One uninfected colony was divided into two; one of these (1, o—o) was given combs removed from colonies found infected with *M. mellificae* in April; the other half (2, ●—●) was given similar combs that had been disinfected with acetic acid (Chapter 10, IV. B.1). B. Colony 1 transferred to uncontaminated combs. C. Colony 2 transferred to comb removed in May from Colony 1. Each point represents the number of infected individuals in a sample of 100 bees. (Partly from Bailey, 1955d).

B. Multiplication

Cysts ingested by the adult bee presumably germinate within the intestine, possibly at the posterior end of the ventriculus where solid food particles accumulate. The infective amoebae then probably enter directly into the Malpighian tubules, which discharge into the posterior end of the ventriculus, and apply themselves to the tubule epithelium. When it excysts the amoeba has a flagellated form that makes its way into the tubule where it changes into the trophic amoeba (Schulz-Langner, 1958). According to Fyg (1932), cysts form in bees kept at 30°C between 22 and 24 days after bees become infected, and then pass into the rectum to be discharged with the faeces.

C. Spread

In temperate climates there is a sharp peak of infection about May in the Northern hemisphere, followed by an abrupt decline, with infection becoming almost undetectable after midsummer (Schulz-Langner, 1969; Hassanein, 1952; Prell, 1926; Poltev, 1953) (Fig. 17). The reason is the same as for the spontaneous decline of infection by *Nosema apis*. When combs of an infected colony are replaced by non-contaminated ones in summer, the infection does not reappear the next year (Fig. 17). Cysts disappear in summer much more abruptly than spores of *Nosema apis*, probably because cysts then have insufficient time to form in the short-lived bees. Natural transmission of infection to the winter bees is almost certainly by the remains of faecal contamination deposited on combs during the preceding late winter and spring. The apparent absence of infection in autumn and winter (Fig. 17) after contaminated combs are introduced, is in marked contrast to the surge of infection by *Nosema apis* when this is introduced in the same manner (Fig. 16). Possibly the development of cysts is more retarded than that of spores of *N. apis* in bees at low temperatures. This could also explain the very steep rise of infection with *M. mellificae* in early spring, when the temperature of the cluster rises from about 20°C to about 30°C (Simpson, 1950).

M. mellificae is associated with *N. apis* in bee colonies more frequently than can be accounted for by chance (Bailey, 1968). The two parasites are independent, since they often occur alone, either in individual bees or even in colonies, but they become associated because they are transmitted in the same way. *M. mellificae* forms only about 500 000 cysts per bee and these taken about 3 weeks to develop, whereas *N. apis* forms up to 30 million spores per bee in about half the time. Therefore, *M. mellificae* spreads less easily than *N. apis*, usually only by severe dysentery. Accordingly, it is usually associated with the most severe infections by *N. apis* (Fig. 18) and with unusual mortality of colonies, but it is not a prime cause of such losses.

Fyg (1932) was unable to infect 7 queens with *M. mellificae* by feeding them cysts in syrup, although the queens were successfully infected with *Nosema apis*, spores of which were fed at the same time; and worker bees were successfully infected with both organisms. Fyg suggested that queens are resistant to infection, as they may be to the pyloric flagellates (Section IV.), but at least one naturally infected queen has been found (Ingold, personal communication).

Örösi-Pal (1963) failed to detect cysts in over 500 queens that seemed to be sick or that were surplus to requirements, but 3 out of 10 queens that he tried to keep in cages in the laboratory during the winter became infected. The attendant workers in the cages were changed every week or so and, although they were not checked for infection, they sometimes defaecated "excessively" and so would have transmitted any infection they had to the queens. Queens become infected rarely in nature, probably because their chances of

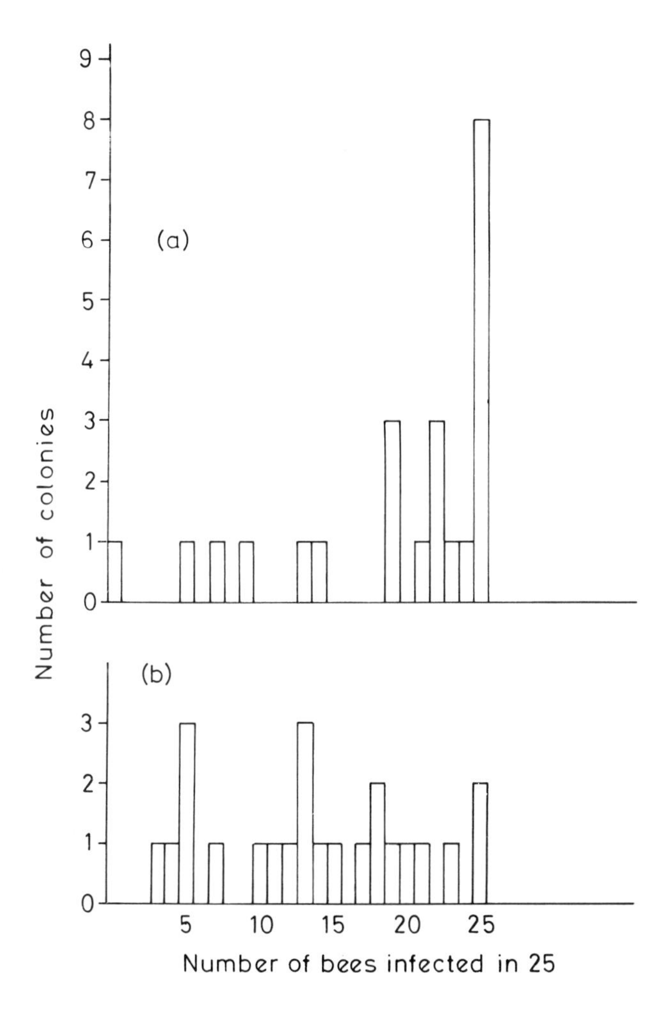

Fig. 18 Distribution of infection with *Nosema apis* (a) and *Malpighamoeba mellificae* (b) among colonies infected with both parasites (compare with Fig. 14). (From Bailey, 1968).

ingesting cysts are fewer even than those they have of ingesting spores of *N. apis* (Section I.C.).

D. Occurrence

Malpighamoeba mellificae has been reported from most European countries, Russia, U.S.A. (Bulger, 1928), New Zealand (Palmer-Jones, 1949) and South America (Stejskal, 1958). However, its incidence usually seems very low, e.g. 2% of colonies in England and Wales (Min. of Agric. Fish. and Food, 1956 and 1959), 0·2% in Italy (Giavarini, 1956) and none in Scotland (Murray, 1952).

III. GREGARINES

These large sporozoa (*Leidyana* and other spp.) live in the lumen of the mid-gut of adult bees attached to the epithelium (Fig. 34e) and may not be specific to honey bees (Wallace, 1966). Similar specimens have been found in Orthoptera and Lepidoptera. Those in bees may originate in nectar-gathering insects or in insects that sometimes infest honeycombs, such as cockroaches, which live in weak colonies in tropical areas (Hitchcock, 1948; Stejskal, 1955). Stejskal (1973) identified four species of gregarines in bees in Venezuela. Gregarines encyst in bees and are passed out via the rectum, but it is not known whether the cysts then mature to form infective spores as they do to complete their life cycle in other insects.

There is little evidence of harm done to bees by gregarines. Infection found in spring in one colony disappeared during the summer (Hitchcock, 1948). Sick bees were collected in October from a few colonies in Venezuela, each bee having about 3000 gregarines in its ventriculus, and the following June sick bees were found once more. However, as Stejskal (1955) reported that fewer than 50% of the sick bees were infected with gregarines, the observed disease could not have been caused entirely by the organisms. Harry (1970) found that gregarines caused no damage to the gut of the desert locust and had no effect on the females, although they caused the weight of males to decrease.

Gregarines have been found in bees in Switzerland (Morgenthaler, 1926), Italy (Giavarini, 1937), Canada (Fantham *et al.*, 1941), U.S.A. (Hitchcock, 1948; Oertel, 1965) and South America (Stejskal, 1955), but their incidence is probably very low. Oertel reported less than 5% of bees infected in Louisiana.

IV. FLAGELLATES

A small area of the epithelium on the dorsal side of the pylorus of adult bees is frequently colonized by protozoa tentatively named *Leptomonas apis* (Lotmar, 1946), but reidentified as *Crithidia* spp. by Lom (1964). Infection causes a dark spot or crust, easily visible on the

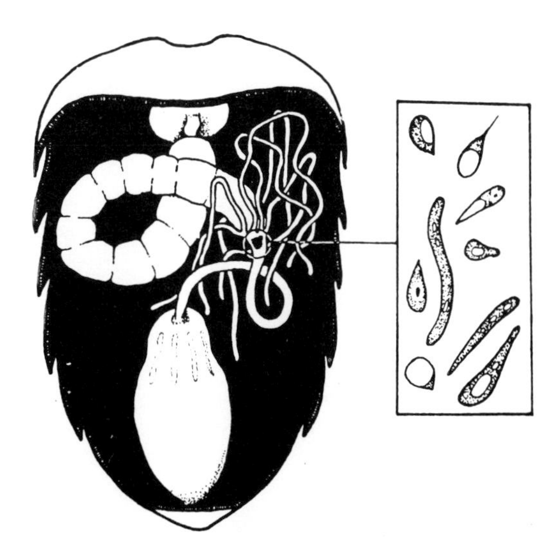

Fig. 19 Flagellates in adult honey bees. The dark spot within the pyloris (ringed) is caused by the accumulation of flagellates illustrated on the right. (After Lotmar, 1946).

intestinal wall (Fig. 19), on which the flagellates are attached to form a "furry" coating. Rounded forms of these protozoa occur and may be cyst stages. Only 1 of 16 queens examined by Lotmar was infected, whereas most of the worker bees examined were infected. Flagellates apparently of the same type, also occur in the rectum, either free in the lumen or attached to the epithelium (Fyg, 1954).

The flagellates do not occur in bees less than 6 days old. In newly-infected bees they move freely in the gut lumen, later they clump into rosettes and adhere to the intestinal wall where the dark crusts appear in bees more than 16 days old. They are scarce in winter bees; sometimes bees with the crusts have been found in winter, but with no flagellates (Giavarini, 1950).

There is no evidence that the flagellates are pathogenic. They have been found commonly in Europe and Scandinavia (Lotmar, 1946) and in Australia, where they were named *Crithidia mellificae* (Langridge and McGhee 1967). They will multiply on artificial media (Fyg, 1954). They may not be specific to honey bees (Wallace, 1966).

7. PARASITIC MITES

There are many different kinds of mite (Acari) that parasitize a wide range of insects as well as other animals. They live and multiply on almost any surface, external and internal, of their hosts; and frequently become highly specific to a certain region, usually a particular groove, cranny or pit from which they are not easily dislodged. Some, for example, spend their lives in the tympanic organs of certain moths, using special sites within the organ for specific activities such as feeding, moulting, and breeding; amazingly, never infesting both ears of the same moth and so not incapacitating their host (Treat, 1958). Many species have taken to living within the tracheal system of insects, including locusts and bumble-bees. Examples such as these serve to show that the several kinds of similar mites that commonly infest honey bees are not so remarkable as they seemed when first discovered.

I. *ACARAPIS WOODI*

This mite (Fig. 20) first called *Tarsonemus woodi* (Rennie, White and Harvey, 1921) was later renamed *Acarapis woodi* (Hirst, 1921). It infests mainly the tracheae that lead from the first pair of thoracic spiracles of adult bees (Figs 21, 35), but mites have also been found in air sacs in the head and abdomen (Prell, 1927).

When congo red is injected into the haemolymph of infested bees, both adult and larval mites quickly turn red; so it appears they feed on the haemolymph of their host by piercing the tracheal walls with their mouth-parts (Örösi-Pal, 1934).

A. Symptoms and Diagnosis

There are no outward signs of infestation by *Acarapis woodi.* Irregular dark stains develop in infested tracheae, the whole of which eventually blacken in infestations of long duration (Fig. 35b). Such deeply stained tracheae seem brittle compared with normal ones but, apart from these effects, little or no internal damage has been detected in infested bees.

The thoracic tracheae leading from the first thoracic spiracles are the main ducts for air, at least into the thorax, and they undoubtedly supply oxygen to the flight muscles. It has frequently been assumed that numerous mites in these ducts partially suffocate the bee and at least impair its ability to fly. However, this is not supported by direct observation. Bees severely infested with mites forage for pollen and nectar in an apparently normal way (Rennie, White and Harvey, 1921); and the same proportion of individuals with severely infested tracheae, some with both tracheae infested, occur among flying bees and those from the cluster of the same infested colonies (Bailey, 1958).

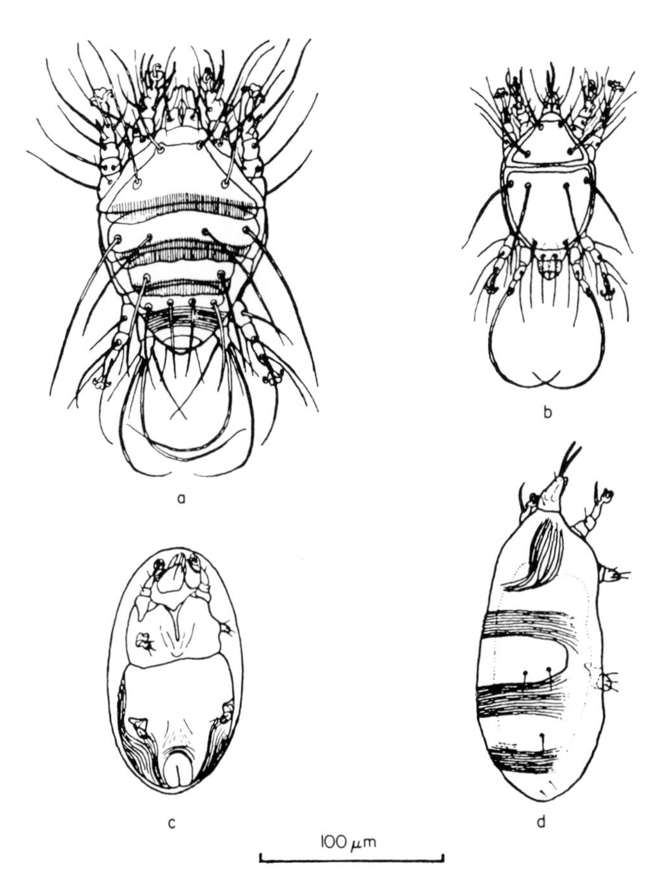

Fig. 20 *Acarapis woodi*: (a) mature female, (b) mature male, (c) larva within its eggshell, (d) mature larva. Sizes vary, but males are always smaller than females. (After Hirst, 1921).

Reports of severe damage of colonies by *A. woodi* abound in beekeeping journals and have become widely accepted, but no experiments were done to measure the effects of the mite until the late 1950s and early 1960s. It was shown then (Bailey, 1958, 1961a) that overwintered infested bees die sooner than uninfested individuals and the difference became statistically significant from about March onwards (Fig. 22). This was done by marking every bee in moderately infested colonies during the autumn and then sampling the marked bees at intervals during the following winter and spring. A sharp fall in the percentage of infested individuals among the marked bees during the spring, showed that infestation was then shortening the lives of individuals. Overwintering bees are near the end of their lives by then, and are being replaced by new generations, so there were relatively few marked bees left to sample, and the effect of infestation on the colonies was not noticeable. However, when colonies are severely infested, most of their adult bees will die slightly earlier than usual, and this may not be sufficiently counter-balanced by the production of new bees. Thus, severely infested colonies dwindle more than usual; and

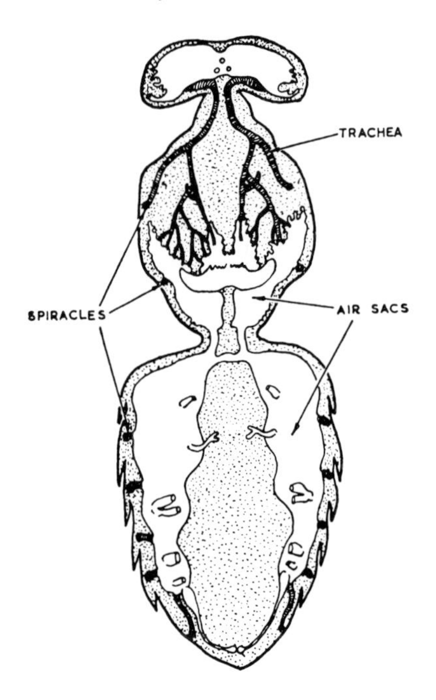

Fig. 21 Tracheae and air-sacs of the adult bee. (After Snodgrass, 1956).

colonies with more than about 30% of their individuals infested are more likely to die in spring than the rest. However, few colonies become so severely infested (Fig. 23), and it has been estimated that in England and Wales, where incidence of *A. woodi* is probably as high as anywhere in the world, something less than 2% of colonies can be expected to suffer measurably from infestation. Most of this small proportion survives, and infestation in most of the survivors diminishes during the active season to the usual small amount.

It was popularly supposed, and is still believed by many, that *A. woodi* causes the crawling and death of bees in summer, leading to severe dwindling of colonies. However, field experiments have shown that infested bees in summer die only a little sooner than uninfested individuals, and they appear normal until they die (Bailey and Lee, 1959).

The common occurrence of *A. woodi* makes it certain that many colonies suffering from any disorder will also be infested to some degree with mites. Some investigations of sick colonies found severely infested in summer with mites showed that, whereas both sick and apparently healthy bees from the colonies were infested about equally with *A. woodi*, the sick bees only were all infected with chronic paralysis virus (Chapter 3, I.A.; Bailey, 1969b). Similarly in northern India, where infestation of *Apis cerana* with mites resembling *A. woodi* is common, a "clustering disease", which leads to the death or severe dwindling of colonies in summer, was attributed to the mites. In fact, the mites were often very few or absent from the diseased bees, but all of these were infected with *Apis* iridescent virus (Chapter 3, II.D.). Many other disorders have similarly been attributed to *Acarapis woodi* in the past, and they have led to the erroneous concept of "Acarine" or "Isle of Wight disease" (Chapter 9, V.A.).

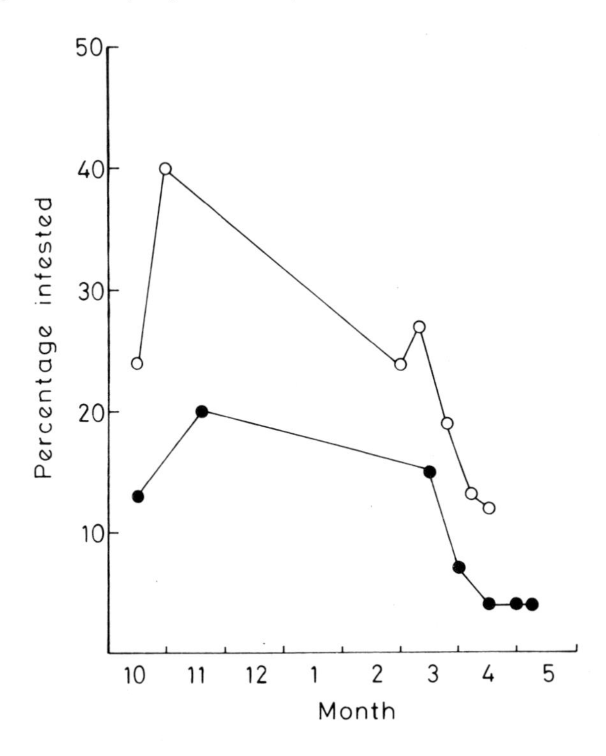

Fig. 22 Percentage of marked individual bees infested with *Acarapis woodi* in 2 colonies, every bee of which was marked in October 1956 (●—●) or October 1957 (o—o). Each point represents the number of infested individuals found in a sample of 100 marked bees. (From Bailey, 1958).

Infested bees have more bacterial infection, particularly in their haemolymph, than uninfested bees (Fekl, 1956), and this may sometimes cause disease. The bacteria presumably invade the haemolymph via wounds made by mites in the tracheae. On the whole the bacteria found in the haemolymph correspond with those in the tracheae, not those elsewhere in the bee. However, infestation with *A. woodi* does not increase the susceptibility

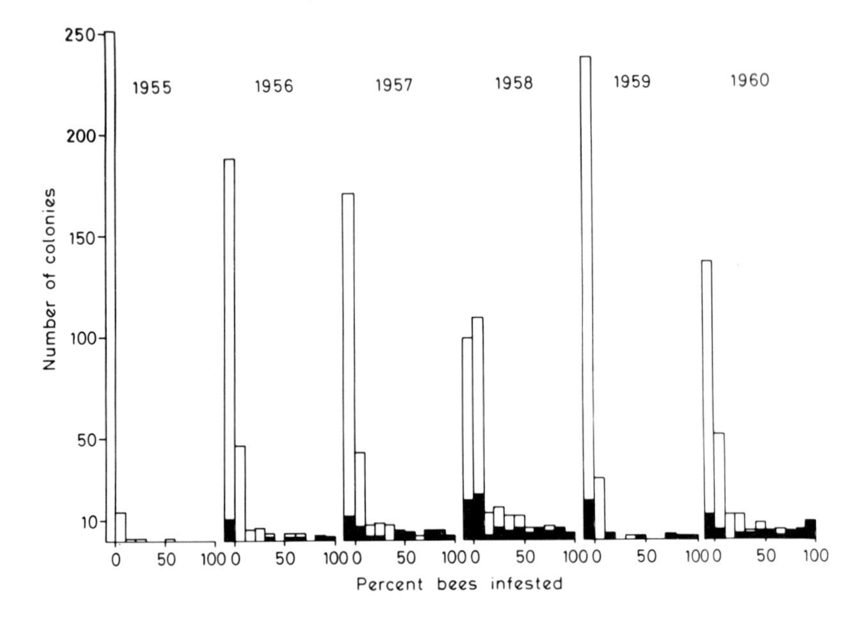

Fig. 23 Natural infestations of untreated honey bee colonies with *Acarapis woodi* in autumn and mortality of colonies in winter at Rothamsted. Numbers of colonies that died during the winters of 1956–1959, inclusive, shown in black. Summers of 1955 and 1959 were the only seasons in the series during which bees accumulated honey surplus to their needs.

of bees to disease when sprayed with suspensions of pathogenic bacteria or of viruses (Bailey, 1965a).

B. Multiplication

Female mites collect within the tracheae of worker bees within 24 h after the bees emerge from their cells and the ratios of male to female mites found within tracheae range from 1:1·1 to 1:3·3 (Morganthaler, 1931a). Single female mites lay 5–7 eggs after 3 or 4 days, and after a further 3 or 4 days the eggs begin to hatch. The first males occur on the eleventh or twelfth day, the first females on the fourteenth or fifteenth day. Other observations have been that the eggs hatch 5 or 6 days after they are laid, and that female larvae become adult mites after a further 6 to 10 days (Örösi-Pal, 1935). Therefore, even when female mites mate and migrate immediately on becoming adult, the youngest worker bee that can possibly transmit mites will be 14 days old.

C. Spread

The proportion of infested bees in any colony often fluctuates widely during short periods

(Fig. 24), which indicates an unstable dynamic equilibrium between the mortality of infested bees and the numbers of healthy bees becoming infested.

Only bees less than 9 days old can become infested (Morgenthaler, 1930, 1931a), and the susceptibility of individuals to infestation diminishes rapidly from their first day of life, although the reason for this is obscure. Some believe it is because the hairs of bees stiffen with age and the dense barrier of hairs at the entrance of the first thoracic spiracles thus becomes impassable (Sachs, 1952); mites can leave the spiracle when migrating, supposedly because the hair-barrier, like a valve, acts only one way. However, when newly emerged and older bees, each with one spiracle denuded of hairs, were introduced into cages of infested bees, some young ones but none of the old ones became infested, and the presence or absence of spiracular hairs had no effect (Lee, 1963).

When a mite leaves a bee via the first thoracic spiracle it climbs a hair, usually on the thorax, and clings near its tip with one or both hind legs (Fig. 35d). It grasps with its forelegs a hair of another bee brushing past and descends to the surface of the new bee's body. Mites are then attracted to the region of the first thoracic spiracle by vibration of the wing roots nearby, and then to the spiracular openings by the puffs of air coming out of them, which are caused by the respiratory movements of the abdomen (Sachs, 1952).

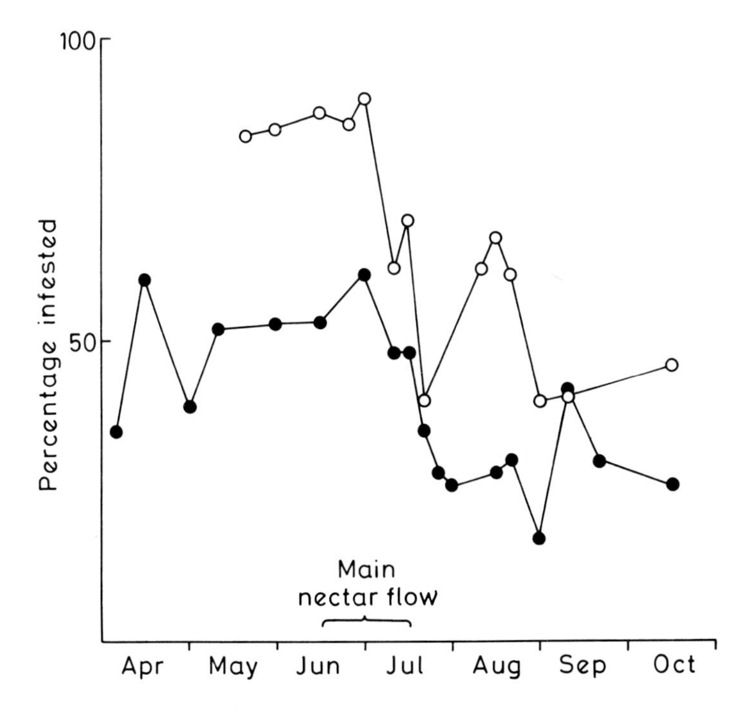

Fig. 24 Percentage of individuals infested with *Acarapis woodi* in 2 individual bee colonies; each point represents number of infested individuals found in a sample of 100 bees (From Bailey, 1958).

Mites migrate only in this fashion: they are unable to find a new host via honeycomb or flowers (Hirschfelder, 1952). They will not even pass through a wire-gauze screen separating infested bees from young uninfested individuals, even though the bees are able to feed one another through the gauze (Morgenthaler, 1931a).

Infestation increases in an endemically infested colony when relatively few young bees are available, because there are then many migrating mites for each available host. It also increases when the foraging activity of old bees becomes suppressed, because there is then more chance of contact between old infested bees and young susceptible ones. These two situations occur together to a large extent; depressed brood-rearing coincides on the whole with times of decreased foraging activity. When foraging increases during nectar-flows, infestation decreases because the old infested bees become separated from the young susceptible individuals (Figs 23, 24).

Plenty of evidence has been gathered to show the marked influence of the type of season on infestation (Fig. 23). However, even in poor seasons, the natural suppression of infestation remains strong and relatively few colonies become severely infested.

In winter, when there are very few young bees and there is an almost static population of insusceptible bees, infestation remains constant, but decreases towards the end of winter as the old infested bees begin to die off a little sooner than the uninfested individuals (Fig. 22).

D. Occurrence

A. woodi was found in Switzerland, France and Czechoslovakia soon after it was found in Britain. Later it was found in bees in: Argentina, Austria, Belgium, Brazil, Chile, Colombia, Corsica, Germany, Holland, Ireland, Italy, Majorca, Poland, Sardinia, Spain, Uruguay, U.S.S.R. and Yugoslavia (Jeffree, 1959; Nascimento, 1971, Menapace and Wilson, 1980); in the Belgian Congo in one of the African bees, *Apis mellifera adansoni* (Benoit, 1959); in Egypt (Bailey, 1981) and in Mexico (Anon., 1981). It seems absent from North America, Scandinavia, Japan, Australia and New Zealand.

As the supposed cause of "Isle of Wight disease" (Chapter 9, V.A.) *A. woodi* was believed by many to have spread from that island. This seems remarkable as there was little or no transportation of bees between most of the regions where the mite was soon found. It seems more likely that it is a long-established and widely distributed parasite of honey bees, but is sparse or absent in certain regions because of environmental factors.

Its occurrence indicates that the mite is potentially ubiquitous in regions where bees are found, but only where there is poor or unreliable forage for bees, the ecological fringes of suitable habitats, is it possible for mites to survive and occur in easily detectable numbers.

In regions with prolonged winters, such as Scandinavia and northern North America, infested bees may not survive long enough to transmit mites to the new generation of bees in spring.

The reason for the apparent absence of mites from temperate and tropical parts of North America, Australia and New Zealand may be as follows: firstly, bees were taken to these parts of the world by man and infested bees may not have survived the long and arduous journeys; secondly, and perhaps more significantly, bees probably survived only in those parts of the world where they could flourish, early settlers would probably not have kept them otherwise. These factors could well have eliminated any infestation that might otherwise have survived the journeys. Even today these regions are more suitable for bees than the areas where *A. woodi* is endemic, and for this reason they may still be largely unsuitable for the survival of the mite. However, there are now many amateur beekeepers in urban areas of North America, Australia and New Zealand, who have bees which are not kept for economic reasons. The food reserves of these bees have to be supplemented with sugar syrup and such relatively inactive colonies could become foci of infestation. Colonies of North American bees imported to Britain were more susceptible to infestation than native bees (Bailey, 1967b), probably because they were less active than local colonies in the British climate. Individually, the American bees were no more susceptible than British bees.

It is significant that the areas where mite infestation is most severe: Britain, Belgium, Switzerland and parts of Germany and Italy; are regions of dense human populations, probably with corresponding densities of their honey bee colonies, many of which cannot be maintained on the available forage.

II. MITES RESEMBLING *A. WOODI* IN *APIS CERANA*

Mites resembling *Acarapis woodi* were detected in specimens of *Apis cerana* from the western Himalayas by Milne (1957) who reported that no morphological difference could be detected between them and *A. woodi* by taxonomists. This, and the possibility that the mites from *A. indica* infest *A. mellifera* under experimental conditions (Atwal, 1967), does not prove their identity, but many biologists have assumed that the mites from *A. cerana* are of *A. woodi*. Moreover, they further assumed that they had been derived from *A. mellifera* that had been taken to India and that they were the cause of severe losses of bees (Kshirsagar, 1966). As described earlier (Chapter 3, II.D.) many of these losses are caused by *Apis* iridescent virus.

III. *ACARAPIS* SPECIES THAT LIVE EXTERNALLY ON ADULT BEES.

There are mites that spend their lives entirely on the outer surface of the bodies of bees but are morphologically almost identical with *A. woodi*. These mites were first found in Switzerland and the first species named was *Acarapis externus*. This species is localized in the area behind the head capsule of adult bees on the ventral side of the neck (Homann, 1933). External mites were then found on bees in Britain (Morison, 1931), but these were localized to the V-shaped groove between the mesoscutum and mesoscutellum (Fig. 25)

Eggs, eggshells, larvae and larval skins usually lie in a contiguous row, with their long axes parallel to that of the groove and generally in its posterior region. These are of *Acarapis dorsalis.*

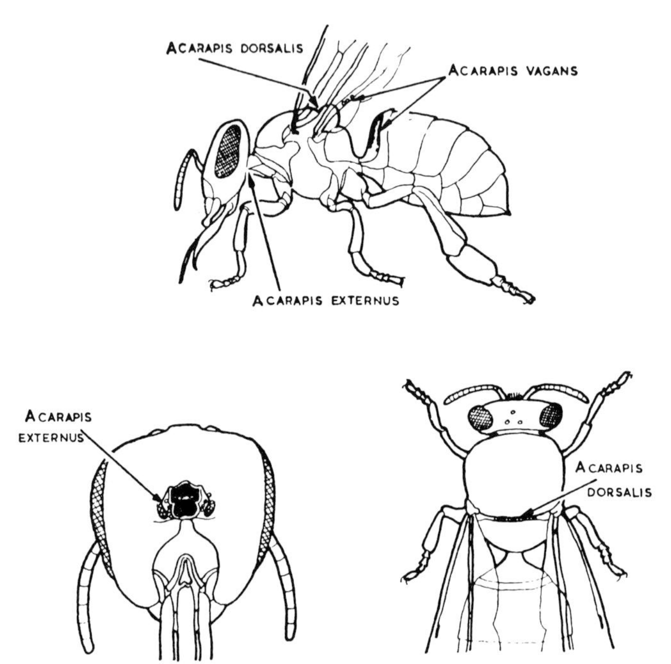

Fig. 25 Location of species of *Acarapis* that live on external surface of adult bees. Lower left-hand diagram shows a rear view of the head. (After Morgenthaler).

The evidence of several workers suggests constant biometrical differences between the three species of *Acarapis* (Brugger, 1936). The length of the terminal segments of the fourth leg of female mites of *A. woodi* and of *A. dorsalis* is between 7·3 and 7·8 μm but that of *A. externus* is about 11·8 μm; so *A. externus* seems well differentiated on this basis. The distance across the body between the two spiracles of the mite is about 13·7 μm in *A. woodi* and about 15–17 μm in *A. dorsalis* and *A. externus*. Although it may be justifiable to distinguish between the mites by these measurements, some taxonomists have considered them insufficient for classifying the mites as separate species. Nevertheless, from the ecological aspect the mites are separate even if they cannot be separated satisfactorily on biometrical grounds. For example, in colonies infested with *A. externus* and observed for 5 years, mites were never found in the tracheae (Morison, 1931); and in more than 46 000 bees from colonies "heavily infested" with external mites, none were found in the tracheae (Borchert, 1929). No correlation has been found between the proportions of bees infested with different types of mite in colonies infested with all three varieties; and colonies have been found infested with two or only one of the three varieties (Brugger,

1936). All these observations strongly suggest that the mites are specifically associated with the restricted localities on the body of the bee where they are found, and that they are distinct species.

A third species of external mite, *Acarapis vagans*, has been proposed. It lives primarily near the roots of the hind wings of bees (Fig. 25); but in severe infestations the forewing, propodeum and first abdominal segment are also occupied, and they spread all over the body of drones. *A. vagans* seems well differentiated from *A. woodi* on biometrical grounds, but it is not clearly separated from *A. externus* or *A. dorsalis* (Schneider, 1941).

There is little information about the life cycle of the external mites. Eggs of *A. externus* are laid on bees about the second day after they emerge in severely infested colonies. The eggs hatch after 4 days and male larvae become adult after 2 or 3 days and female larvae after 4 days. The whole life cycle, therefore, takes about 10 days for a female, which is about 4 days less than that of *A. woodi* (Brügger, 1936). All stages of *A. dorsalis* die on dead bees after about 3 days (Morison, 1931). External mites apparently feed on the haemolymph of adult bees, like *A. woodi*, since they take up congo red that has been injected into the bodies of infested bees (Örösi-Pal, 1934). Their pathogenicity for bees has never been investigated.

Examination of dead bees in winter and flying bees in summer, both from the same colonies, has shown that the percentage of bees infested with *A. dorsalis* and *A. externus* is highest in winter and lowest about midsummer (Brügger, 1936; Homann, 1933; Clinch, 1976). The youngest bees are the most susceptible. Of 40 newly emerged bees introduced to a colony, 31 were found infested with *A. externus* 2 weeks later, but only 6 out of 20 bees introduced when 4 days old and 1 out of 17 bees introduced when 6 days old became infested (Brügger, 1936). Colonies severely infested with *A. vagans* were found to be mostly queenless (Schneider, 1941); and few external mites occur in normal colonies. In all these respects external mites resemble *A. woodi*.

Morgenthaler (1930) found that *A. externus* occurs throughout Switzerland, whereas *A. woodi* is more localized. *A. dorsalis* is commoner than *A. externus* although colonies with *A. externus* usually have a higher percentage of infested bees than colonies with *A. dorsalis*. External mites have been found in Scandinavia, North America, Australia and New Zealand (countries where *A. woodi* has not been found) as well as in Europe, U.S.S.R., South America and Africa. There is no obvious explanation for the wider geographical distribution of *A. dorsalis* and, to a lesser extent, of *A. externus*, than that of *A. woodi*. There is some evidence that *A. dorsalis* is less restricted to young bees when it migrates, and it may breed more profusely in winter than *A. woodi*, which might account for its wide occurrence (Brügger, 1936). *A. externus* seems to have a shorter life cycle than *A. woodi*, which may help it to survive better than *A. woodi* by enabling it to migrate sooner to young bees.

IV. *VARROA JACOBSONI*

This mite (Fig. 35f) was described by Oudemans (1904) when it was first recognized in the brood cells of *Apis cerana* in Java. It is the only parasite of honey bees that can readily be seen with the naked eye and be identified with a hand lens.

The female mite, which is a dark red-brown, lays up to twelve eggs in a brood cell, preferring that of a drone, just before it is capped. The nymphal stages of the mite feed on the haemolymph of the immature bee and may kill it. Otherwise, and more usually, the mites attach themselves to the emerging bees. These sometimes have deformed wings. The attached mites are mature females and are already fertilized. Male mites, which are smaller and paler than the female, die shortly after mating within the sealed brood cells. The females continue to feed on the haemolymph of adult bees, usually attached between the overlapping segments beneath the abdomen, but they eventually migrate to larvae about to be sealed in their cells. In summer they can live for about 2 months; later they can live for up to 8 months and so can survive the winter with the hibernating cluster of bees when there is little or no brood.

Mites that die, or are killed by treatments (Chapter 10, V.B.), fall off the bees and can be detected on the surface of sheets of white paper, or similar material, inserted onto the floor of the hive. These inserts are best left for several weeks, or throughout the winter. Mites can be separated from wax particles and other rubbish by extraction in alcohol, on which they float (Ruttner and Ritter, 1980).

V. jacobsoni is common on *A. cerana* in East and South East Asia, but it became notorious only after it was found in *Apis mellifera* in Eastern Europe, West Germany, North Africa, the Middle East and South America (Crane, 1978). Apparently the mite was taken on live queens of *Apis mellifera*, or on the accompanying workers, from Eastern U.S.S.R. to European Russia, and similarly from there to Europe and the Middle East. It is alleged to have been carried in the same way from Japan to South America.

Theoretically, the mites can reproduce and spread fairly quickly, multiplying perhaps five-to ten-fold every 2 weeks, and there are many popular accounts of their propensity to overwhelm and destroy colonies. However, they are alleged not to cause obvious damage to colonies for 3 or 4 years, or even longer, after they become established (Ruttner and Ritter, 1980). Evidently, their ability to multiply and spread is counterbalanced by their mortality and by their losses on adult bees that die in the field. Severe infestations probably occur in colonies that, for reasons other than infestation with the mite, suffer unusual setbacks in brood-rearing and foraging activities. Such events could well affect the average colony every few years, especially in relatively poor regions for bees. Then, the females of *V. jacobsoni* attached to adult bees would have the greatest opportunity to invade the relatively few brood cells, just as *Acarapis woodi* invades more young bees than usual in the same circumstances. This, in fact, is the underlying principle of the manipulative method (Chapter 10, V.B.) developed by Ruttner and Ritter (1980) for trapping and removing migrating female mites from colonies.

It is notable that the presence of the mite excites little concern among beekeepers in regions where bees usually flourish, such as parts of South America, and doubts have been expressed about its alleged virulence even in Germany and east Europe. No good evidence about its effects on colonies are forthcoming from Japan and Asia, where it has probably been established in *Apis mellifera* ever since this species of bee was taken there. In fact, the mite was brought to European Russia from the eastern U.S.S.R. on strains of *A. mellifera* which had attracted attention because of their high productivity (Crane, 1978).

Conceivably, *V. jacobsoni* could act as a vector for certain pathogens that commonly occur as inapparent infections but are very virulent when injected in the haemolymph. Acute bee-paralysis virus (Chapter 3, I.G.) is an example, and it has been associated with *V. jacobsoni* and severe losses of bees in the U.S.S.R..

Many chemical treatments that have been recommended are harmful to bees (Chapter 10, V.B.) and the damage they have done to bee colonies since they began to be used from about 1965 has probably been attributed to *Varroa jacobsoni*.

V. OTHER MITES

Several other mite species that live externally on the brood or adults of honey bees have been reported from the Far East and South East Asia. The commonest is *Tropilaelaps clareae* (Delfinado and Baker, 1961) (Fig. 35e), which has been found infesting adults, pupae and even dead individuals of *Apis mellifera* and *Apis dorsata* in India, the Philippines, Vietnam and Hong Kong (Crane, 1968). Curiously enough, it has not been found in colonies of *Apis cerana*. Its effect on colonies is unknown. It has been found on sick bees in weak colonies but it may not be the primary agent of such damage. It becomes more numerous in queenless than in normal colonies (Laigo and Morse, 1968), which suggests it is usually naturally suppressed in ordinary circumstances, in ways similar to those that suppress the other mites of bees.

8. INSECT AND NEMATODE PARASITES

I. DIPTERA

A. Larvae

The larvae of some species of fly parasitize adult honey bees, which are then said to be suffering from "apimyiasis". The adult female fly of *Senotainia tricuspis*, a common European species, waits near entrances of beehives and swoops down on flying bees to deposit a tiny, newly-hatched larva on their backs, usually on the joint between head and thorax. Flies may contain 700–800 larvae and can deposit 1 per bee every 6–10 seconds (Boiko, 1958). The flies are not specifically associated with honey bees: two species of *Bombus* were found parasitized in an area where 40% of honeybees were infected (Boiko, 1948).

The larva is armed with piercing mandibles which enable it to penetrate the intersegmental membrane where it is deposited, and to enter the haemolymph of its host. It grows in the abdomen or, more usually, the thorax of the bee, but does not feed on the solid tissue while its host is alive (Simmintzis and Fiasson, 1949).

When the bee dies, the fly larva eats the thoracic muscles, moves to the abdomen and eats the soft tissue there, and finally leaves the shell of its host to pupate. If it encounters another dead bee, however, it will eat the bee's contents first and then pupate. Larvae do not enter living bees after leaving their dead host (Giordani, 1955).

Larvae pupate in the soil and emerge as adults after 7–16 days; or in the following year, about July in the northern hemisphere, after a winter period of diapause.

Lesions to thoracic muscles of bees parasitized by *S. tricuspis* and *Melaloncha ronnai*, a Brazilian species, have been reported, but abdominal myiasis caused by either fly seemed to cause no damage (Guilhon, 1950). *M. ronnai* has been said to paralyse its host in which it also pupates (Ronna, 1936); but it is a member of the Phoridae family, and these "hump-backed flies" are generally scavengers on dead bees (Milne, 1951). They parasitize moribund bees, and Guilhon (1950) described "pseudo-myiasis" caused by post-mortem infestation by numerous species of Diptera. However, Örösi-Pal (1938a) cites examples of them being true parasites. *Borophaga incrassata*, another of the Phoridae, has been said to lay eggs in bee larvae that continued to develop and even became sealed over, often with their heads to the bases of their cells, and then died (Paillot, *et al.* 1944). Severely affected brood has been described as resembling foul-brood or chilled brood.

Drosophila busckii has been found in the thoraces of a few bees (Mages, 1956). The fly resembles *S. tricuspis* but apparently develops entirely within the bee, metamorphosing within the thorax or the abdomen.

Peak infection by *S. tricuspis* occurs in August in southern France, and only hives exposed to bright sunlight are affected (Simmintzis and Fiasson, 1951). But the proportion of parasitized bees usually seems low: out of 40 000 bees examined in July, August and September, only 69 were found parasitized (Rousseau, 1953).

There are very conflicting statements about damage to colonies by apimyiasis. *Sarcophaga surrubea* has been considered damaging to honey bees in North and South America (Braun, 1957). *S. tricuspis*, when first discovered in 1929 by Angelloz-Nicoud and for many years afterwards, was thought to cause devastating losses: inability to fly, paralysis and crawling in front of the hive, all being given as signs of infection (Simmintzis, 1958). Methods of control were advocated, legislation contemplated and even enforced in some European countries. Severe damage by *S. tricuspis* has been said to occur in the U.S.S.R., with 70 or 80% of bees lost from colonies in the Ukraine and "mass mortality" near the lower Dnieper (Sukhoruka, 1975). Yet others report no sign of damage, even in colonies with 80% of bees infected (Giordani, 1955).

Diseases caused by pathogens that are less readily detected than fly larvae are probably often mistaken for apimyiasis, especially when many bees are parasitized.

S. tricuspis has been reported mainly from France and Italy. It also occurs in North Africa (Mathis, 1957) and Australia (Roff, 1960). *Rondaniooestrus apivorus* causes apimyiasis in South Africa (Milne, 1951).

B. Adults

Braula coeca is associated specifically with honey bees. Eggs (Figs 26, 35h) are laid on the inner side of the cappings, and sometimes the walls, of cells full of honey. Subsequent development is entirely beneath the cappings of honey cells and not among brood cells. The grub-like larva (Fig. 26) makes a tunnel of wax fragments which it gnaws from the cappings. This tunnel, traversing several cells, is very thin at first but expands as the larva grows. Larvae presumably obtain protein food from pollen which, together with wax, occurs in their intestines. Micro-organisms within the epithelial cells of the mid-intestine may help to digest wax. When fully grown, the larva pupates within its tunnel. The adult emerges and finds its way to the bodies of worker or queen bees and occasionally to drones. Often many collect on the queen at once, probably because she is the most permanent member of the colony rather than especially attractive. Daily collections from a single queen have totalled 371, about 30 being found at any one time. *B. coeca* feeds by moving from its usual place on the constriction between the thorax and abdomen of the bee to the head when its host is about to feed, and takes up a position on the open mandibles and labium. It reaches into the cavity at the base of the extended glossa of the bee near to the opening of the duct of the salivary glands to obtain food, possibly salivary gland secretion (Imms, 1942).

B. coeca is usually described as an inquiline, and therefore not harmful, but it is possible that severe infestations decrease the efficiency of queens.

Örösi-Pal (1966) mentions several species and sub-species of *Braula* and describes three groups represented by (1) *B. coeca,* (2) *B. schmitzi* and (3) *B. pretoriensis*, the latter being the type in Africa.

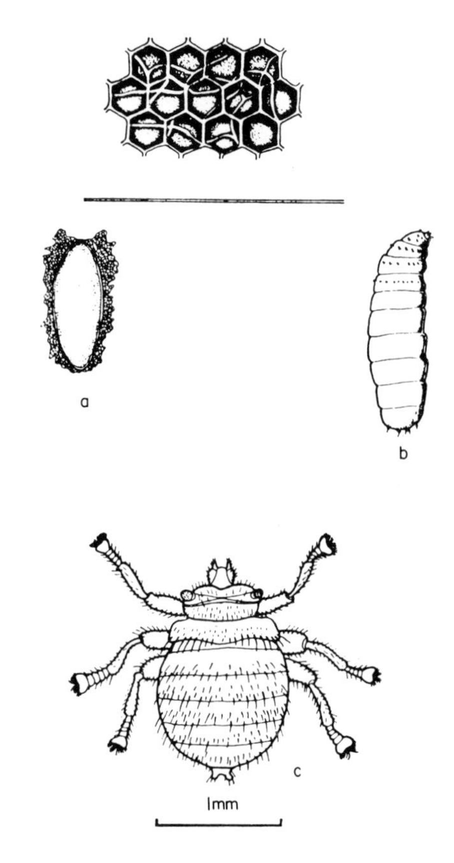

Fig. 26 *Braula coeca*. Above: underside of honeycomb cappings showing wax tunnels of larvae of *Braula coeca*; below: (a) egg, (b) larva (c) adult. (Partly after Imms).

II. COLEOPTERA

Larvae of beetles (*Melöe* spp.), first called *Pediculus apis* by Linnaeus and now known as triungulin larvae, hatch from their eggs on the soil, climb onto open flowers and attach themselves to the bodies of aculeate Hymenoptera. The active larva, about 1 mm long, drops on to the egg in the nest of solitary bees to be sealed up with the store of pollen and honey. It eats the egg, changes to the grub-like larvae and then, after eating the stored food, pupates and emerges as an adult oil beetle a month later or more usually after hibernating through the winter.

Larvae of *M. cicatricosus* have been said to find and eat the eggs of honey bees, to turn into grubs that look like bee larvae and, astonishingly, to be fed by the adult bees (Seltner, 1950).

Triungulin larvae have been reported to burrow in the joints of the abdomen and thoraces of adult bees, which then die with convulsive movements. Hundreds of such bees have been seen dying on the ground of Russian apiaries (Beljavsky, 1933).

Larvae of *M. variegatus* on bees injected with congo red took up the dye, which suggests that they pierce the body wall and suck the blood of the bee (Örösi-Pal, 1937a). *M. proscarabaeus*, when tested in the same way, did not take up the dye but were seen to consume honey from the comb.

M. variegatus occurs in North America, Russia, Europe and Britain, but *M. proscarabaeus* is the common species in Britain and does not seem to harm bees.

III. STREPSIPTERA ("STYLOPS")

Species of Strepsiptera have been found parasitizing honey bees (Johnsen, 1953). The insects have initial larval forms that look like triungulins of oil beetles, but they parasitize larvae of various species of Hymenoptera, living in their body cavities in a grub-like form and feeding on the haemolymph which diffuses through their body walls. The larvae continue to develop and emerge as adults, although they may be deformed: the pollen-collecting apparatus may be absent or the sting may be diminished in size.

Female stylops live permanently within their host, their head and genital apparatus protruding through the bee's cuticle. Male stylops are free-flying and mate with the female after alighting on the host insect. Ulrich (1964) doubts that stylops can complete their life cycle in honey bees, although the larvae are probably carried to the colony by foragers. Certainly, honey bees are rarely parasitized by stylops and damage is probably slight.

IV. LEPIDOPTERA

Wax moths usually destroy combs unoccupied by bees, but in warm climates the greater wax moth, *Galleria mellonella* often overwhelms small colonies. Occasionally, the lesser wax moth, *Achroia grisella*, and, less often, *G. mellonella*, chew away some of the cappings of sealed brood and the adult bees remove the rest to cause a condition known as bald-brood. The absence of cappings over pupae is of little consequence, but some of the pupae emerge with deformed legs and wings (Milne, 1942).

V. NEMATODES

Mermithid nematodes have occasionally been found in honey bee workers, drones and queens. They are 10–20 mm in length, very thin whitish worms that parasitize many insect species. Mature adult worms dwell in the soil and mate there. The eggs are laid on wet grass by the active female, or else the young larvae make their way there. Insects eat the grass, or take dew from it, and the eggs are ingested or the young larval nematodes penetrate the insect cuticle.

Queen honey bees probably receive nematode eggs that are brought in by bees collecting water. One queen whose ovaries contained no eggs was found to have an encapsulated nematode in her body cavity near the hind-gut and ovary (Kramer, 1902).

Nematodes probably have to go through a soil phase, so they are not likely to multiply and spread within bee colonies. Their natural hosts are probably ground-dwelling insects, including solitary bees and bumble-bees. However, Vasliadi (1970) reported some 60% of many hundreds of honey bees he examined to be parasitized by mermithids in low-lying regions of the USSR.

Mermis nigriscens has been found in honey bees in Switzerland and an *Agamermes* spp. was found in honey bees in Brazil (Toumanoff, 1951). *Mermis albicans* has been found in worker, queen and drone honey bees in Europe (Fyg, 1939; Paillot *et al.* 1944) and larval mermithids were found in worker bees in eastern USA by Morse (1955).

PLATE SECTION:
Figures 27 to 35

Fig. 27 (a) Comb of bees and capped brood (Courtesy of Dr Ingrid H. Williams); (b) cages for laboratory tests, each suitable for up to 40 adult bees; S, W = vials containing strong sucrose solution and water respectively; P = cork with holes for pollen taken from bee combs.

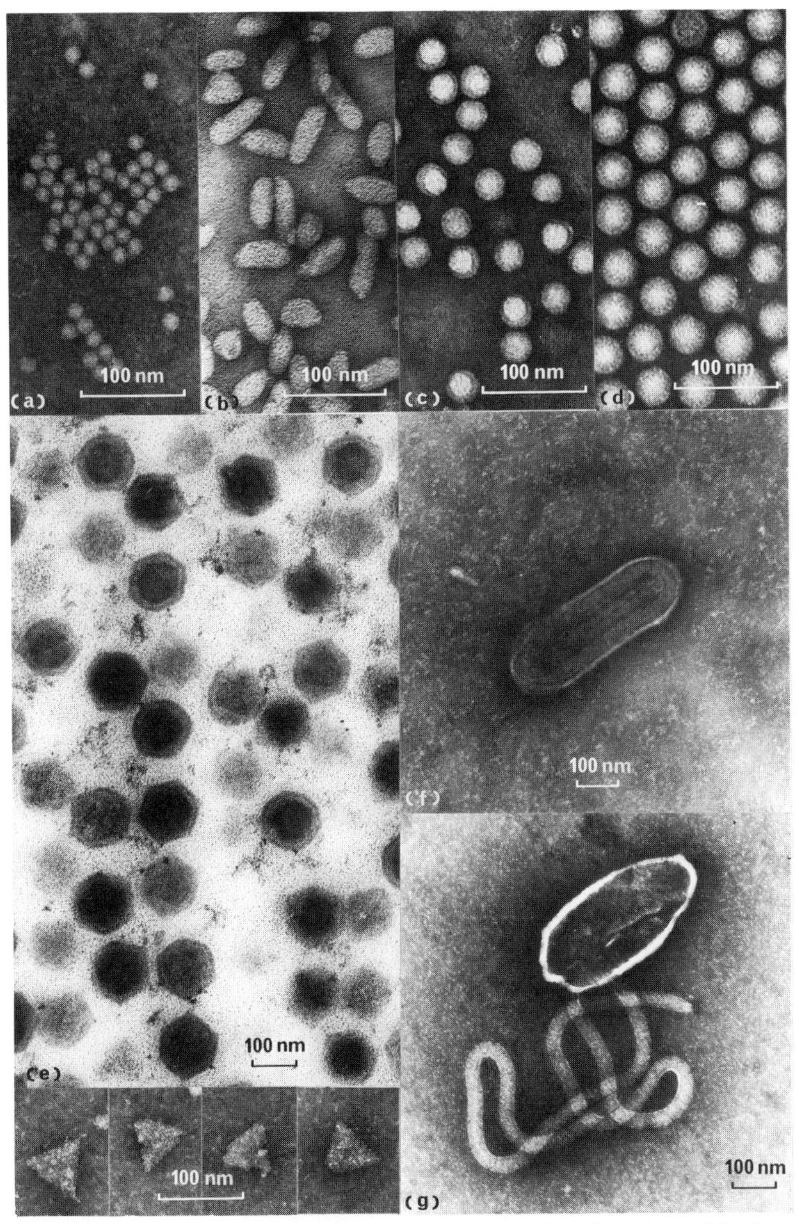

Fig. 28 Electron micrographs of representative types of viruses from bees. (a) particles 17 nm in diameter (cloudy-wing particle, chronic bee-paralysis virus associate); (b) chronic bee-paralysis virus; (c) particles 30 nm in diameter (sacbrood, black queen-cell, acute bee-paralysis, Kashmir bee, Egypt bee, slow paralysis and Arkansas bee viruses); (d) particles 35 nm in diameter (Bee viruses X and Y); (e) *Apis* iridescent virus in ultrathin section of cytoplasm of adult fat-body cells; lower inset: trisymmetrons of sub-units that form the outer shell of the virus particles; (f) filamentous virus particle; (g) filamentous virus particle with ruptured envelope releasing the single flexuous rod, or nucleocapsid, which contains DNA and measures 3000 nm × 40 nm. All stained with sodium phosphotungstate except (f) which was stained with ammonium molybdate.

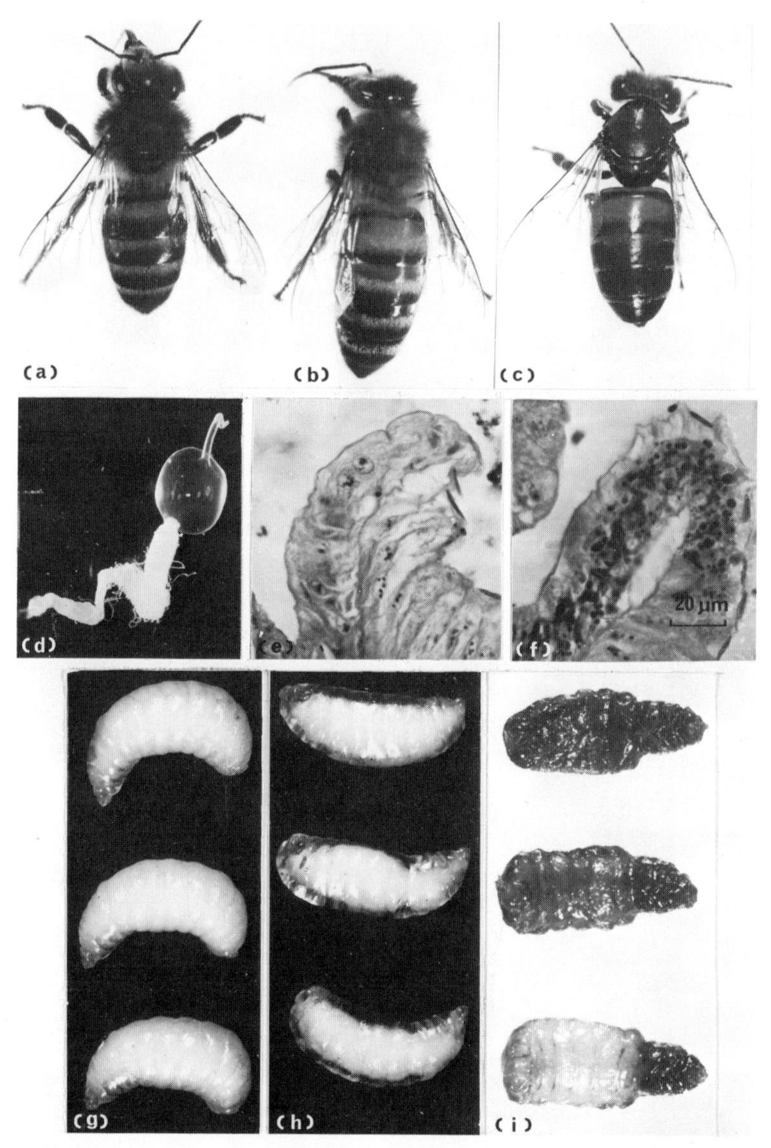

Fig. 29 Chronic bee-paralysis: (a) healthy individual; (b) Type 1 syndrome; (c) Type 2 syndrome; (d) gut of bee with Type 1 syndrome; the bloated honey-sac causes distension of the abdomen (see (b)); (e,f) longitudinal sections of gut epithelium immediately posterior to the Malpighian tubules of a healthy bee and of a bee with chronic paralysis, showing Morison's cell inclusions in the latter (Heidenhain's iron haematoxylin).
Sacbrood: (g) healthy individuals; (h) early stage of disease; (i) formation of scale (bottom to top).

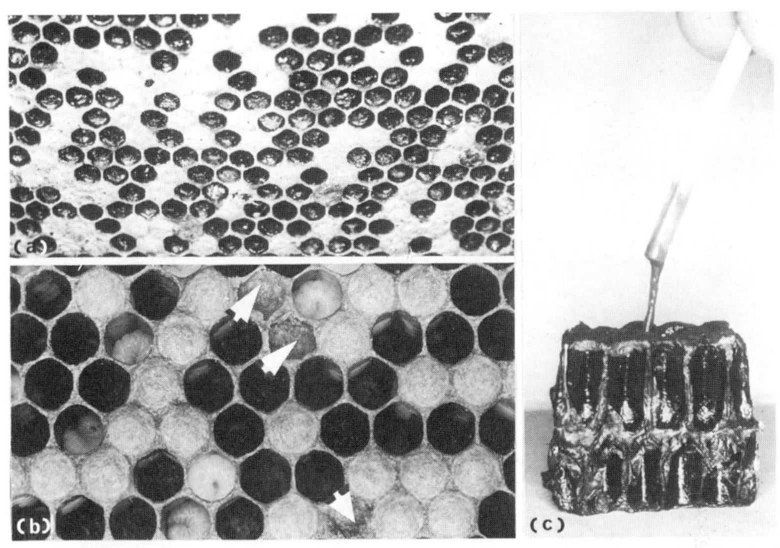

Fig. 30 American foulbrood: (a) oblique view from above of comb with remains ("scales") of severely attacked brood; (b) dark cappings (arrowed) of dead pupae among healthy brood; (c) ropy thread formed with larval remains.

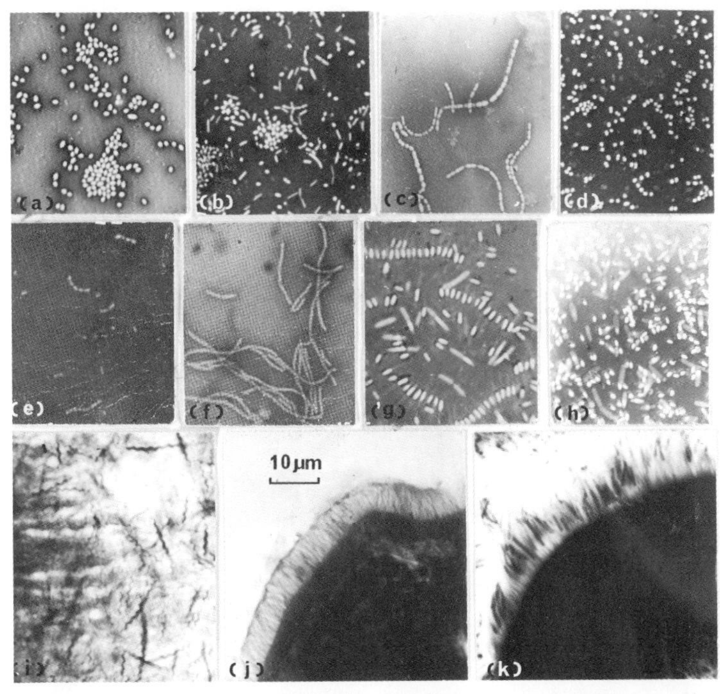

Fig. 31 Bacteria of European and American foulbroods (×650): (a) *Streptococcus pluton* from larvae; (b) *Streptococcus pluton* + "Bacterium eurydice" from larvae; (c) *Streptococcus pluton* from culture; (d) *Streptococcus faecalis* from culture; (e,f) "Bacterium eurydice": coccal and rod forms of one strain cultivated on pollen-extract- and honey-based media respectively; (g) *Bacillus alvei* spores; (h) *Bacillus larvae:* sporulating culture; (i) *Bacillus larvae:* coalesced flagella in gut contents of a young larva; (j) Brush border of the mid-gut cells of a healthy larva; (k) *Bacilllus larvae:* vegetative cells in the brush border of the mid-gut cells of a young larva.

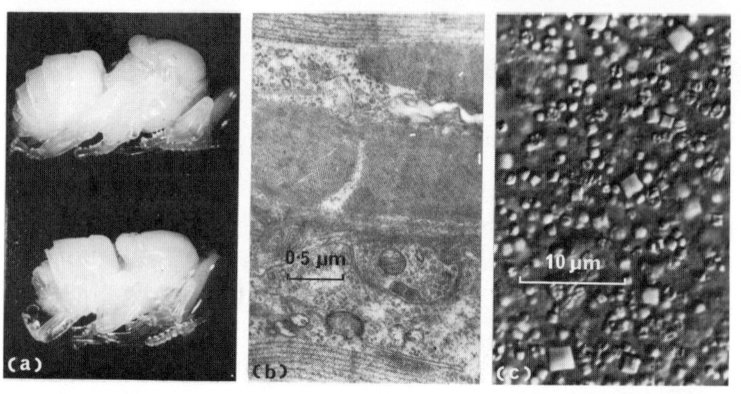

Fig. 32 (a) Healthy pupa (top) and a pupa from a larva that had been infected with *Streptococcus pluton*; (b) ultrathin section of thoracic muscle of an adult bee infected with cloudy-wing particle showing (top to bottom) muscle fibril, sarcosome, tracheole, crystal of particles, neuromuscular junction, sarcoplasm, muscle fibril; (c) cubic bodies of a polyhedrosis virus that attacks wax-moths.

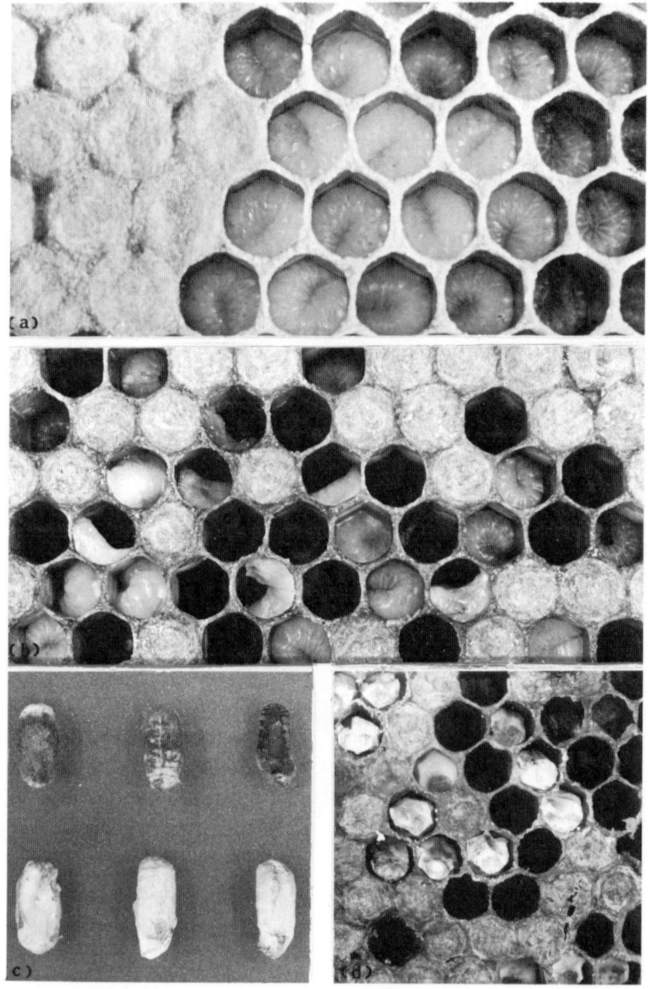

Fig. 33 European foulbrood: (a) healthy larvae, capped larvae on the left; (b) larvae with European foulbrood.
Chalkbrood: (c) top row: prepupae covered with fruiting bodies of *Ascosphaera apis*; bottom row: prepupae killed by a single strain of *A. apis*; (d) severely attacked brood.

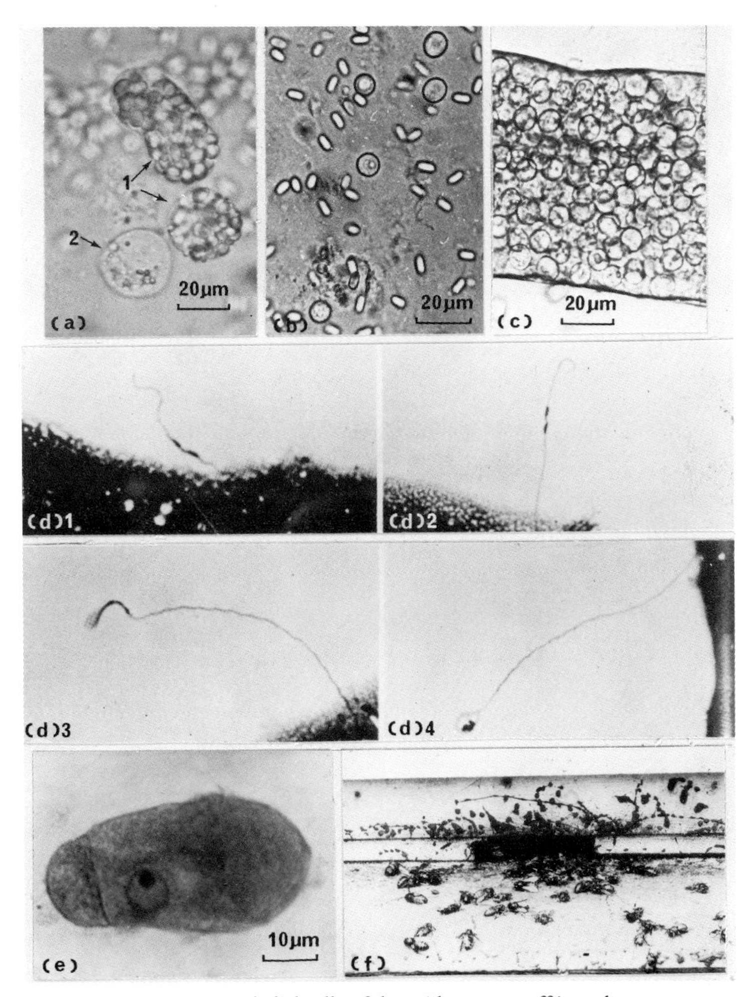

Fig. 34 (a) *Nosema apis*. Live epithelial cells of the mid-gut cast off into the gut contents of infected bees, 1. cells packed with spores, 2. uninfected cell; (b) spores of *N. apis* and cysts of *Malpighamoeba mellificae*; (c) cysts of *Malpighamoeba mellificae* in a section of the Malpighian tubule of an adult bee; (d) filaments of *N. apis* extruding from hanging drops, showing the passage of the twin nuclei down the hollow filament (1, 2) and their emergence in the sporoplasm (3, 4), which in nature is injected into the host cell (Courtesy of J. P. Kramer); (e) gregarine from the mid-gut of an adult bee (Courtesy of J. D. Hitchcock); (f) "Dysentery": photograph taken in 1911 of entrance of a bee colony alleged to have "Isle of Wight disease"; appearance of dead bees at hive entrance and of faecal matter on hive parts are not uncommon after long winter confinement of normal bee colonies (Courtesy of the Bee Research Association).

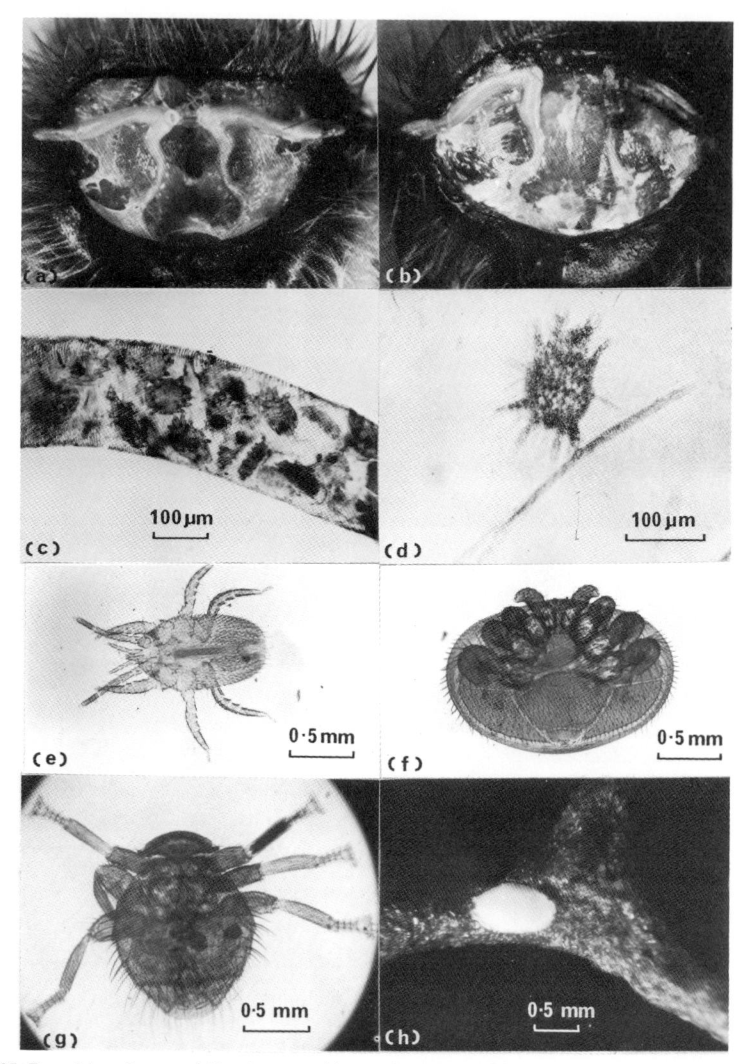

Fig. 35 Parasitic mites, and *Braula coeca*; (a) anterior view of thoracic tracheae of healthy adult bee after removal of head and first thoracic segment; (b) trachaea blackened unilaterally as a result of infestation with *Acarapis woodi*; (c) tracheae filled with individuals of *A. woodi*; (d) Migrating female of *A. woodi* gripping a hair of its old host with one leg and ready to attach itself to the hair of a passing new host; (e) *Tropilaelaps clareae*; (f) *Varroa jacobsoni* (female); (g) adult *Braula coeca*; (h) egg of *Braula coeca* on the edge of a honeycomb cell ((a, b and c) courtesy of H.M.S.O.).

9. DISORDERS OF UNCERTAIN ORIGIN AND NON-INFECTIOUS DISEASES

Several non-infectious disorders are often confused with infectious diseases. Dysentery and various kinds of poisoning are the commonest.

I. DYSENTERY AND POISONOUS SUGARS

By common usage among beekeepers, "dysentery" means defaecation by adult bees within or near the colony so that the comb and the hive entrance are visibly soiled with excrement (Fig. 34F). Bees accumulate material in their rectums when they are prevented from flying, so defaecation within the hive occurs mainly in late winter. Usually it is slight and of little consequence, but sometimes it is severe and associated with the rapid death of affected colonies. The bees die from a combination of suffocation and turmoil in their efforts to clean themselves and their combs of the excrement, and of the infectious diseases that it often carries.

Langstroth (1866) noticed that bees confined to their hives in damp conditions were apt to become dysenteric; he also considered that feeding bees on liquid honey in autumn or disturbing bees in winter were the prime causes of dysentery. Dadant (1890) and Root (1901) thought that dysentery was caused by dilute honey, fruit juices that beekeepers sometimes fed to bees, or by honey-dew, which is the excrement of aphids often collected by bees; and Phillips (1945) considered dextrins, which occur in honey-dew, caused dysentery. However, Lotmar (1934) showed that bees can digest dextrins as well as starch partly decomposed by boiling, and Alfonsus (1935) demonstrated that dysentery was not caused during winter when colonies were fed during the preceding autumn either on 15% dextrin in sucrose, honey-dew, strong sucrose solutions, or large amounts of pollen in sugar candy. He showed that dysentery was caused by feeding liquid honey, or dilute or crystallized sucrose solutions. Dysentery occurs when the rectal contents of bees become about 30–40% of their total body weight (Alfonsus, 1935, Lotmar, 1951) and this is due to the accumulation of water, usually because there is too much water in the food.

Dysentery often occurs when honey stores become coarsely granulated: the bees cannot ingest the large crystals and the liquid contains too much water. Sometimes this dilute honey ferments and, although the yeasts and alcohols produced by the fermentation are probably harmless, the process produces even more water.

However, there is an important kind of dysentery, which, although ultimately the result of water accumulating in the rectum, is not because the food contains too much water, but because it contains certain toxins. The toxins, as yet unidentified chemically, are produced by the action of organic or inorganic acids on glucose or fructose, and they appear to upset the normal metabolic water balance within the bees. They shorten the lives of bees by 50% or more when they are the only source of carbohydrate food. The most striking sources of these toxins are acid-inverted sucrose and acid-hydrolysed starch which have frequently been marketed as sugars suitable for feeding to bees (Bailey, 1966).

Hydrolysed sucrose is commonly, but mistakenly, believed to be nutritionally better for bees than sucrose because of its similarity to honey, and it is also favoured because it can be used to make a candy that stays soft and suitable for feeding to bees. However, unless hydrolysed with enzymes, it quickly causes dysentery and kills bees. Honey that has been heated is similarly toxic for bees, because of the action of its natural acids on the glucose and fructose. Even honey that has been stored at ambient temperatures for several years causes dysentery and shortens the lives of bees compared with those fed on sucrose. Sucrose partially hydrolysed by boiling with cream of tartar or vinegar, recommended by many beekeeping manuals as the basis of a solid candy for feeding the bees, is also toxic for bees when compared with plain sucrose.

In view of these facts and of the toxicity of unrefined sugars, including semi-refined beet sugar, the only form of sugar that can be recommended unreservedly for feeding to bees is refined sucrose. Moreover, concentrated sucrose syrup is far less likely to crystallize after it has been stored in the comb by bees than natural honey, because it contains a much lower proportion of glucose (Simpson, 1952).

An extreme example of crystallization of honey was given by Greenway *et al.* (1975) who observed that honey largely composed of ivy nectar contained 70% glucose, whereas the average concentration of glucose in honey is only 40–50%. Because of its high glucose content, the ivy honey granulated quickly and then lost water by evaporation, unlike ordinary honey, which is hygroscopic even when granulated, because of its high fructose content. Many colonies in Eire that had stores largely composed of ivy honey, died during the winter, apparently because they could not obtain sufficient water from their food.

II. PLANT POISONS

There seem to be several sources of poisonous nectars and pollens (Burnside and Vansell, 1936; Maurizio, 1945), although the poisoning of bees by them is not easy to verify, especially as incidents are frequently local and transient.

In the U.S.A. trouble arises from California buckeye (*Aesculus californica*). It has been estimated that 15 million acres of foraging area becomes virtually unavailable for bees for 6 weeks in early summer when buckeye, a species of horse-chestnut, is in bloom. Both pollen and nectar are injurious; saponin is believed to be the toxin. Adult bees become trembly and lose their hair, partly or wholly, because unaffected bees pull at them as if trying to eject them from the hive (cf. Paralysis, Chapter 3, I.A.1.). Young brood are killed, deformed pupae arise from sealed brood and queens are said to become drone-layers. The nectar of horse-chestnut contains 50–60% sugar and is therefore very attractive to bees and, since the tree is resistant to drought, it may at times be virtually the only source of nectar.

Honey and pollen from lime (*Tilia* spp.) may be toxic. Many hairless, dark, trembly honey bees (cf. Paralysis, Chapter 3, I.A.1.) together with solitary and bumble-bees with the same appearance, have been seen under lime trees in midsummer, and the honey bees seemed to have no known infectious disease according to Maurizio (1943), although bee viruses were virtually unknown at the time.

Honey-dew from lime is sometimes toxic, and its high content of the trisaccharide melezitose has been blamed, although bees are known to be able to digest the sugar; and honey-dew from *Tilia platyphylla*, which killed bees within 8 h, contained a soluble, heat-stable toxin (Butler, 1943), unlikely to have been melezitose. Possibly it was galactose and/or mannose, both of which occur in the nectar and honey-dew of lime and are toxic for bees (Barker, 1978). Honey-dew from conifers is said to poison bees sometimes, particularly in Europe, where the disease is called *Waldtracht-krankheit*. However, samples of bees suffering from this disease have been found to be severely affected with chronic paralysis virus (Bailey, 1976; Chapter 3, I.A.1.).

The nectars of *Rhododendrum ponticum* L. and other *Rhododendrum* species contain andromedotoxin, which is poisonous for bees. Some species of *Ranunculus* (buttercups) have anemonal in their pollen which is also poisonous to bees. Buttercup pollen may be collected, particularly when cold weather delays cherries and dandelions, and resultant damage to bees has been reported from Switzerland where it is called "Bettlach May disease".

Astragalus lentiginosus (spotted loco weed) is troublesome in the U.S.A. at times: pupae die and mummify, adults turn black and hairless, and queens are said to die frequently from poisoning. *Astragalus* species can accumulate toxic levels of selenium, but toxicity to bees may not be related to this (Barker, 1978).

Cyrilla racemiflora (southern leatherwood) is said to cause "purple brood" in southern United States: unsealed larvae turn purple and die in early summer.

In experiments, nectar or pollen from death camas (*Zygadenus venenosus*) killed honey bees within 2 days, but strangely enough the solitary bee *Andrena zygadeni* appears to forage exclusively on death camas blossom (Hitchcock, 1959).

There seems little doubt that there are sources of natural poisons for honey bees, although many causes of natural poisoning are often difficult to establish. In view of more recent knowledge of hitherto unknown virus diseases of bees (see Chapter 3), reports of bee poisoning from plants should be treated with reserve.

The most common feature of natural poisoning, whatever the origin, is that more usual sources of nectar or pollen have failed, usually because of drought or cold.

III. INSECTICIDES

Poisoning of bees with insecticides has become an important hazard, sometimes causing severe losses which may be confused with the effects of infectious diseases of adult bees and larvae. Fortunately, most manufacturers and users of insecticides are aware of the danger to beneficial insects, and efforts are made increasingly to develop insecticides that are selective for pests, either by reason of their chemical nature or by their method of application (Wilson *et al.,* 1980). An assessment of the relative toxicities to honey bees of a wide range of insecticides has been given by Stevenson *et al.* (1977).

IV. HEREDITARY FAULTS

Eggs that fail to hatch, larvae that fail to pupate and pupae that die of no apparent infection are sometimes described as "addled" and are believed to be suffering from hereditary faults.

About half the number of eggs from greatly inbred queens do not produce larvae, but disappear to give areas of unevenly developing brood. Mackensen (1951) predicted and demonstrated this. He believed that the eggs that disappeared were of diploid drones, i.e. had arisen from fertilized eggs, instead of being the usual haploid drones characteristic of all Hymenoptera. All the fertilised eggs were laid in worker cells, but half of them were believed to be homozygous at a sex-determining locus of a chromosome and therefore male, and they were believed not to be viable. This mechanism of sex-determination and the non-viability of diploid drones had been found earlier in the wasp *Habrobracon juglandis*. The fact that diploid male tissue could exist in honey bees was later demonstrated by Rothenbuhler (1957) in gynandromorph or mosaic bees. Then, Woyke (1962, 1963, 1965) successfully hatched in an incubator all the eggs from inbred queens that had each been mated to one drone and that were laying eggs of which only 50% seemed to hatch in colonies; and he reared them to maturity in the laboratory. Half were drones and, as they reflected the genotype of their father, they were presumed to be from fertilized eggs. Later, Woyke and Knytel (1966) showed that such drones indeed had the diploid complement of 32 chromosomes. Therefore, the larvae of diploid drones are viable but, as Woyke (1963) showed, they are detected in bee colonies as soon as they hatch and are immediately eaten by nurse bees.

Instances of queens producing eggs which all fail to hatch (Hitchcock, 1956) are inexplicable on the genetic grounds given above and the fault may not be hereditary. Most cases of "addled brood" described by Tarr (1937d) in Britain were sacbrood (Chapter 3, I.C.3.), once believed not to exist in Britain but now known to be very common. "Addled brood" due to hereditary faults is probably less common than was once believed.

V. DISEASES OF UNKNOWN ORIGIN

A. The Isle of Wight Disease

The Isle of Wight disease was alleged to affect adult bees and was said to have reached epidemic proportions in the British Isles on at least three occasions between 1905 and 1919. The main symptom usually given was very many bees crawling and dying on the ground outside their hives (Fig. 34f). The first major outbreak was said to have been on the Isle of Wight in 1906. The disease was then believed to have spread to the mainland in the south of England in 1909 and, according to Herrod-Hempsall (1937), by 1918 "not a beekeeping district in Great Britain was free from scourge [and] . . . eventually the parasite [*Acarapis woodi*] invaded Ireland as well as European countries". This is the common belief, and it is typified by a statement about *Acarapis woodi* issued by the American Beekeeping Federation which says: "This Isle of Wight disease is considered by apiculturists in the countries where it does exist to be far more serious than American foulbrood".

There is no doubt that some beekeepers lost most of their bees in the Isle of Wight in 1906, which, apparently, was the worst of two or three consecutive bad years. It was then assumed, however, without any evidence, that the cause of the losses was an infectious disease. This idea was then promulgated by sensational but uninformative articles, in *The Standard*, a now defunct London morning paper, and in several provincial newspapers. The publicity, as usual, helped to fix the belief firmly in the public mind.

The first professional investigation was made by Imms (1907). He examined bees on the Isle of Wight which were said to have the disease and found they had "enlargement of the hind intestine" which Imms, who at the time seemed unfamiliar with bees, thought abnormal. His diagram, however, represents very clearly the intestine of any normal bee that has been long confined to the hive. Malden (1909), the next professional investigator to visit the Isle of Wight, pointed out that the intestines of healthy bees confined to hives for a few days very closely resembled those of diseased bees. He had accepted the idea that there was an infectious disease, however, and he obtained a colony, said to have Isle of Wight disease, and confined them in a "warm room" in a muslin cage on 17 June 1908. By 10 August, he said, they had ceased to fly; and the colony was dead by 26 October. To keep them for so long under such conditions, however, would have been difficult had the colony started in the best of health.

Malden examined minutely the anatomy of bees said to have the Isle of Wight disease, including their tracheae and air sacs, but all he found were more bacteria in the intestines of diseased bees than in those of healthy ones; he failed to show that these micro-organisms were pathogenic. Bullamore (1922) also pointed out that bees prevented from flying sometimes develop symptoms, described as crawling with bowel distension, which were indistinguishable from those alleged to be of the Isle of Wight disease. In 1906, according to newspaper accounts, there was a disastrous April for agriculture, with frost (−5°C in London on 2 May) and snow after a very early spring, which had been hot enough to draw crowds to the seaside resorts. This very unusual weather might have accounted for trouble with bees, which, being suddenly confined to their hives, possibly with freshly gathered nectar, may well have become very dysenteric (Section I.).

There is no more recorded evidence about the disease in those early days in Britain. There are, however, descriptions of the death of numerous colonies between 1901 and 1905 in several other countries, including Italy, Brazil, Canada and the United States, and all the bees had symptoms exactly like those described for the Isle of Wight disease in Britain (Bullamore, 1922). One incident was the loss of 20 000 colonies in Utah, with the bees dropping to the ground, mounting blades of grass and twigs with great difficulty and then dying. Had this disaster occurred in Britain at the same time it would have been attributed unreservedly to the Isle of Wight disease. Subsequently, there have been further reports of large scale losses of bees in many parts of the world, particularly in Australia and South America, with bees crawling and dying in front of their hives and with no known parasite present in sufficient numbers to be the cause. Poisonous nectar or pollen was suspected, but the causes may well have been viruses (Chapter 3, IA, IIA). Whatever their causes these losses would certainly have been classified as the Isle of Wight disease by beekeepers in Britain 50 years ago.

There are all kinds of possible reasons for the death of bees, apart from infections, and there is little doubt that bees dying of non-infectious diseases were often included in casualties attributed to the Isle of Wight disease. Imms (1907) found the most successful remedy was "feeding cane sugar" and in Cumberland, where the disease was said to be serious in 1915 and 1916, with between 5 and 20% of colonies "affected" according to a report of their beekeepers' association at the time, it was said that "1916 was a poor season, many colonies were insufficiently provided for winter, and sugar was practically unobtainable". It appears therefore, that starvation was often to blame for some losses included in casualties alleged to be due to the Isle of Wight disease.

So-called treatments for the disease must have killed numerous colonies. One official report said diseased bees were short of nitrogen, because their distended rectums contained much pollen. This followed the mistaken belief that adult bees usually did not need protein food and, when they did, that pollen was unsuitable; so it was recommended that all pollen combs should be removed in autumn and the colonies fed beef extract to make good their supposed nitrogen deficiency. This would certainly kill or seriously cripple any colony because pollen is essential for adult bees and beef extract is poisonous for them, mainly because of its salt content. The ruinous idea of removing so called "pollen-clogged" combs persisted, however, and was widely practised for many years. Other remedies that were recommended were phenol, formalin, "Izal", sour milk, salt and other chemicals lethal to bees, all to be fed in syrup to ailing colonies and as preventives to healthy ones. Other reports describe colonies which clearly were crippled with foulbrood; and poison sprays were certainly used, probably with less consideration than today for bees. Perusal of all the British bee journals from their beginnings until about the 1920s, shows that many beekeepers eventually attributed all colony deaths that had no obvious cause to the Isle of Wight disease. Some beekeepers were sceptical; they pointed out that the symptoms were not specific; e.g., they resembled those of the fairly well-known disease called paralysis, for which there was no known cause, but which had been described from time to time at least half a century before the Isle of Wight disease. The final opinion of Rennie (1923), a co-discoverer of *Acarapis woodi*, who had much experience with bees said to have the Isle of Wight disease, was that "under the original and now quite properly discarded designation 'Isle of Wight disease' were included several maladies having analogous superficial symptoms".

The publicity had won long before, however: by about 1912 almost everyone had accepted the idea that the Isle of Wight disease was infectious and thought that only the identity of the one supposed infectious agent was needed. This encouraged a burst of activity which culminated in the discovery of *Acarapis woodi* in December 1919 by Rennie *et al.* (1921) who at first considered this mite to be the cause. Their own results did not support this, however; on the contrary they made it clear that *A. woodi* was widespread, occurring in many apparently normal colonies. Their report showed that many bees from both diseased and healthy stocks behaved and flew normally, even though they were infested with mites, some of them with pronounced blackening and hardening of their infested tracheae. Normal nectar- and pollen-gathering bees from stocks in which "crawling and other symptoms were well established" were found heavily infested "quite as badly as anything . . . observed in crawling bees". In fact "flying workers were frequently more heavily parasitized than were bees of the same stock which were unable to fly". This evidence shows that *A. woodi* was not obviously pathogenic and certainly could not have been causing the observed sickness which was considered to be the Isle of Wight disease. It seems the mite was then much as it is today: its only significant pathological effect being to shorten very slightly the life of bees, but usually causing no obvious sickness in spite of the abnormal appearance of infested tracheae (Chapter 7, I.A.). Why *A. woodi* became so firmly established as the cause of the Isle of Wight disease in the face of this evidence is hard to understand. It may have been partly because of the size of the parasite, its incidence, and the appearance of infested tracheae, which were startling; but its restricted habitat in the thoracic tracheae, which are neither easy to see without a special dissection technique, nor of obvious interest, make its late discovery understandable. Perhaps some thought it was the last adult bee parasite that would be found and, as the other parasites known at the time, *Nosema apis* and *Malpighamoeba mellificae*, did not seem very dangerous, it was believed, *A. woodi* must be the cause of Isle of Wight disease. This ignored the possibilities of other pathogens, especially of viruses, which we now know to cause diseases with symptoms resembling those reported to be of Isle of Wight disease.

The confusion of thought about *A. woodi* is illustrated by the account in the book by Herrod-Hempsall (1937), which perhaps best reflects the popular beliefs of those days. He stated that *A. woodi* spread from the Isle of Wight to European countries after 1918. Yet in the same account he wrote "there is little doubt that [*A. woodi*] has infested the honey bee in a number of countries for several centuries". His second statement is nearer the probable truth, which is that the mite has infested honey bees for several thousand millennia; it almost certainly has no other host and it is extremely closely adapted to lead a life intimately associated with honey bees. It is widespread and has been found in India and Africa as well as Europe, Russia and America. It was found in France and Switzerland in the winter of 1921–22 and even in Tula, south of Moscow in 1922 (Perepelova, 1927); which is most remarkable progress if it started from England, especially considering the difficulties there must have been in transporting bees during those times. *A. woodi* is unknown in the United States, Canada, New Zealand and Australia, but so it probably would be in Britain if we enjoyed their comparatively regular and abundant nectar-flows. For it is in these conditions that mites decrease in number, quite possibly to become extinct (Chapter 7, C.). And it is in the opposite circumstances, poor seasons when colonies are having a lean time, that *A. woodi* multiplies and spreads. These were the seasonal conditions in which Rennie and his colleagues discovered the mite, and his colonies which had suffered the worst conditions developed the severest mite infestations. Mite infestations increase in colonies as a result of their poor circumstances: it then adds to these, occasionally being the last straw, but usually it dwindles dramatically when the environmental conditions for bees improve enough to make them forage actively (Chapter 7, I.C.).

To sum up, there is no evidence that any parasite we know today has been the cause of wholesale losses of bees. All the information leads to the suspicion that the Isle of Wight disease was assumed to be the cause of all the losses for which there was no obvious explanation at the time. Beekeepers saw their bees die; they knew that infections caused sickness and death in other animals and, lacking knowledge, yet feeling the need for an explanation, they assumed an infectious disease was killing their bees.

B. Other Undiagnosed Abnormalities

1. *Queen bees*

Fyg (1964) made many meticulous studies of the abnormalities of adult queens in Switzerland. He observed that some queens lay a disproportionate number of drone (unfertilized) eggs for pathological reasons, and not because they are unmated or have exhausted their supply of spermatozoa. They have plenty of spermatozoa, although many of these are usually coiled singly in the spermatheca instead of lying normally in wavy bundles. This abnormal feature of the spermatozoa is not absolutely diagnostic: 17 out of 223 of these "drone-broody" queens had normal sperm. The epithelial cells of the spermatheca always seem to break down, however, and their nuclei contain round or oval inclusion bodies which are slightly refractile, take acid stains and are of varying size and number. Fyg suggested that the inclusion bodies are caused by a virus, but their great variability in size and their staining qualities seem to differentiate them from the inclusion bodies that have been recognized in known virus infections of other insects (Chapter 3). When drone-broody queens lay a great many drone eggs, many of the resulting larvae are neglected, because of the shortage of nurse bees, and they die.

Fyg also describes stones that consist mainly of concentric layers of uric acid, each attached with a stalk of tissue to the rectal epithelium, sometimes occurring in the rectums of queens. They vary in size, usually being about 1 mm in diameter, though some of over 2 mm in length have been found, and they may be yellow, reddish-brown, grey or brownish-black in colour. Bacteria almost always occur on the stalks of the epithelial tissue that lie within the stones. Only a small proportion of queens with stones have other recognized infections, so the cause of the stones remains unknown. In the worst cases the stones seem to exert pressure on the genital duct and hinder oviposition. Stones somewhat smaller than those in queens are also found occasionally in worker bees.

2. *Worker bees*

Various non-infectious abnormalities of worker bees have been recorded. Rare cases have been reported of pupae that die shortly before they are due to emerge with the cuticle of their heads and first pair of legs still unpigmented. Some propupae develop similarly when they are removed from their comb and incubated artificially. The cause is blockage of the first thoracic tracheae with the final moulted cuticle, which is not completely stripped off (Fyg, 1958). This presumably prevents enough oxygen reaching the tissues to allow the cuticle to darken normally, but the reason for the failure to moult properly is unknown. Fyg also found worker pupae with abdomens extraordinarily retracted and sometimes with enormously distended heads. For unknown reasons their honey-stomachs and mid-guts have become enclosed in the thorax and head, and the abdomen is retracted by the hind-gut which is still attached to the last abdominal segment and the mid-gut.

A fibromatous-like tumour, thought to be of neural origin, has been observed in the anterior lower portion of the thorax of a worker bee (White, 1921); and giant vacuolated cells have been seen amongst the epithelial cells of the hind-gut of old overwintered bees (Örösi-Pal, 1937b). The giant cells were formed apparently by the fusion of several epithelial cells and averaged about 100 μm in diameter, but some were up to 0·5 mm long; 18% of bees in one colony were found to be affected.

Dark-brown crusts have been seen in place of contiguous epithelial cells of the mid-gut in adult bees (Giordani, 1956). The bees were partially paralysed but attempts to infect healthy bees with the diseased tissue failed (Amici and Vecchi, 1958).

Wille (1967) describes two degenerative changes, of unknown cause, of the muscle tissues of sick bees. One is associated with "sharp-edged crystalloids" within the degenerating muscle fibres. The changes all frequently occur with other well-known infections, but they are not specifically associated with any of these.

3. *Larvae*

Larvae of all ages sometimes die without any apparent infection. "Chilled brood" is then often diagnosed, the implication being that not enough adult bees were present to cover the brood and keep it warm when they clustered at night or in inclement weather. However, unsealed larvae can survive several days at room temperature (about 18°C) without food, so severe or prolonged chilling would be necessary to kill them.

10. THE TREATMENT OF BEE DISEASES

I. VIRUSES

At present, and in common with nearly all virus diseases of animals, there are no known direct treatments for virus infections of bees. However, it is probable that black queen-cell virus, bee virus Y and filamentous virus will be suppressed indirectly by the control of *Nosema apis*, with which they are intimately associated (Chapter 3, I.D.). Measures that prevent dysentery will also probably prevent the spread of bee virus X (Chapter 3, I.E.).

Beekeepers have long believed that paralysis can be cured by replacing the queen of an affected colony with one from elsewhere. There is much circumstantial evidence to support this belief (Chapter 3, I.A.2.) and a policy of replacing old queens with some from other parts of the country rather than allowing colonies constantly to requeen themselves and become inbred, is probably a good precautionary measure. On the other hand, importing queens from great distances runs the risk of bringing in exotic pathogens, and of introducing strains of bees that are unusually susceptible to local viruses. For example, a significantly greater proportion of colonies headed by queens imported into Britain from Russia developed visible signs of sacbrood than did colonies headed by local queens (Bailey, 1967d). A degree of cross-breeding that would be likely between well dispersed colonies in nature would probably be the ideal compromise.

There is evidence that chronic paralysis, cloudy wing and Kashmir bee viruses are transmitted most readily, either by direct contact or by aerial transmission between individuals when these become unusually crowded. Beekeeping methods that lead to crowding and confinement can be expected to exacerbate these, and possibly other, virus infections.

II. BACTERIA

A. American Foulbrood

1. *Destruction of diseased colonies*

This is best when the incidence of disease is low. The colony is killed with about half a pint of petrol poured in the top of the closed hive, and then burned in a shallow pit which is afterwards filled in. The boxes and equipment are usually flamed thoroughly with a blow torch, but a better treatment is to immerse them for 10 min in paraffin wax heated to 150°C (Cook, 1968). There is no doubt that destruction suppresses the incidence of American foulbrood; but the evidence suggests that it cannot eradicate endemic disease from large areas. For example, a compulsory scheme of destruction that operated in Switzerland from 1908 was followed by a fall in the incidence of American foulbrood from approximately 13 to 2% after 25 years (Leuenberger, 1933). Furthermore, the number of colonies inspected increased over the years and the actual number of those found diseased fell only from about 90 to about 40 or 50 per year at the end of the period. In England and Wales, the number of colonies found and destroyed each year for 20 years since 1954 did not fall appreciably from an average of about 800 out of 80 000 colonies inspected each year (Min. of Agric. Fish. and Food, 1974).

2. Manipulative treatment

Shaking adult bees from diseased colonies on to new combs has often been tried, but the method is unreliable. In one investigation (Winter, 1950) disease incidence could not be decreased below 2%; in a another, 81 colonies out of 300 that were treated redeveloped disease in the following 2 years (Dunham and King, 1934); and many colonies derived from adult bees that were shaken into empty boxes from diseased colonies, fed syrup for 72 h and subsequently transferred to wax comb-foundation became diseased again (Sturtevant, 1933). The method was officially endorsed in Denmark but, after many years, American foulbrood remained the commonest honey bee disease in the country (Kaare, 1952, 1953).

Improved methods of transferring bees have been suggested. Bees can be anaesthetized with smoke from fuel impregnated with potassium nitrate. They then fall off the combs without gorging themselves on honey as they do when they are smoked in the usual way. This method, followed by no feeding for about 24 h, considerably decreases the chance of spores in honey being transferred to the new combs; but it is laborious, and the old combs which have to be destroyed are at least as valuable as the bees.

3. Sterilization of combs and honey

Much effort was devoted to methods of sterilizing combs and equipment in the 1920s and 1930s. Formaldehyde was most frequently used; it was evaporated from its solution (formalin) at normal temperatures, by heat, or by adding potassium permanganate or barium peroxide to formalin in closed boxes containing the combs; alternatively, the combs were soaked in soap–formalin, alcohol–formalin or glycerol–formalin mixtures. All the methods work to a large extent. The simplest is fumigation with vapour from excess formalin at room temperature, and it seems no less efficient than the others. Combs enclosed with about 150 ml of formalin per 25 litres of volume at normal temperatures are mostly completely sterilized after 14 to 17 days; open cells are sterile after 6 days. The main drawback is the difficulty of sterilizing capped cells and the permanent toxicity to bees of honey and pollen in the fumigated combs.

Chlorine, either as a gas or in watery solution, is effective, but it makes combs brittle and severely corrodes metal parts (Hitchcock, 1936). It seems to be no more efficient than formalin.

More recently, interest has revived in methods of sterilizing combs, because their value and their importance in the transmission of disease, together with the shortcomings of chemotherapy, have become better appreciated. The most promising new fumigant seems to be ethylene oxide. Shimanuki, Knox and Herbert (1970) showed that boxes of combs from colonies with American foulbrood were almost completely sterilized after humidification for 24 h and then treatment with ethylene oxide at 1 g/litre for 48 h at 43°C in fumigation chambers. Boxes can be fumigated cheaply, and possibly effectively, in plastic bags filled with a mixture of ethylene oxide (10%) and carbon dioxide for about a week at 27°C (Winston, 1970); but for best results, heated, humidified, gas-tight containers are required (Gochnauer and Corner, 1976). However, ethylene oxide forms an explosive mixture with air and its use, even in gas-tight chambers, does not guarantee sterility (Wilson *et al.*, 1978; Gochnauer *et al.*, 1979).

The fumigation of empty comb and equipment that have been in contact with American foulbrood is probably the safest and most economical practice. Combs containing dead brood are best destroyed by burning.

Although honey can be sterilized by acidification with phosphoric acid followed by heating (White and Sturtevant, 1954) the process makes honey extremely poisonous for bees (Chapter 9, I.).

Combs can be disinfected with gamma irradiation from Cobalt – 60. Gochnauer and Hamilton (1970) found that, although absolute sterility is not readily achieved by 1·0 M rad, infectivity was almost completely eliminated by 0·2 M rad. Such a process could be economical when done on a sufficiently large scale using commercially available facilities.

4. Chemotherapy

Sodium sulphathiazole effectively suppresses American foulbrood when 0·5–1·5 g are given in 5–15 litres of strong sucrose syrup to a diseased colony (Haseman and Childers, 1944). The drug is stable in honey for several years and can be supplied in autumn to act successfully during the following year (Katznelson and Jamieson, 1955).

Oxytetracycline ("Terramycin") also is effective when 0·25–0·4 grams in 5 litres of syrup is supplied to a diseased colony. All its activity disappears after about 2 months in honey but higher concentrations are toxic for bees (Section B.4.).

Other methods of applying drugs have been recommended, principally dusting them over the combs in dry sugar, but at best this is no more effective than the methods described and is more likely to damage brood. Mixing drugs with vegetable fat and sugar, and applying them in so-called "extender-patties" may be more efficient than applying them in syrup in some circumstances (Wilson *et al.*, 1971a).

The efficiency of drug treatment varies widely. The degree of contamination of equipment, the ability of the beekeeper and the variability of the many natural factors that influence the course of disease all affect the issue.

Chemotherapy has no effect on spores that contaminate combs and equipment. Its use can lead to the spread of such infection between colonies and so to an increasing dependence on regular application. Accordingly, chemotherapy is not advisable when the incidence of American foulbrood is low and can be contained readily and economically by the destruction of relatively few colonies. Nevertheless, when applied efficiently, chemotherapy can decrease infection within diseased colonies, sometimes with remarkably little recurrence of disease (Wilson *et al.*, 1971b). Chemotherapy becomes economically attractive when disease is widespread, but it would be advisable to withold treatment after one or two seasons of use to see whether it is needed any longer.

5. Resistant strains of bees

A certain degree of resistance to American foulbrood is common in bees (Chapter 4, I.E.), and tests in the U.S.A. by Park (1935) showed that some colonies were more resistant to American foulbrood than others. The early tests were done with an inoculum per colony of the dried remains of 75 larvae killed by the disease, which is about the critical amount (Chapter 4, I.E.): the resistant colonies were unable to overcome bigger inocula (Eckert, 1941; Filmer, 1943; Woodrow, 1941a, 1941b, 1942). Woodrow and Holst (1942) noticed that many individually infected larvae in the resistant colonies were removed about 6 days after they were sealed in their cells, whereas very few sister larvae placed in control colonies had been removed 11 days after they were sealed, by which time they were full of infective spores. This was confirmed by Rothenbuhler (1958) and his colleagues who made further selections for resistance and susceptibility, increasing both about ten-fold compared with the average.

Rothenbuhler and his associates went on to demonstrate a range of hereditary factors that contributed towards resistance. These comprised

(1) the efficiency of the hygienic behaviour of adults in removing diseased larvae, which was further separable into a factor for uncapping the cells and a factor for removing the larvae (Rothenbuhler, 1964);

(2) the rate at which young larvae became innately resistant to infection with increasing age (Bamrick and Rothenbuhler, 1961);

(3) the efficiency of adults in filtering the spores of *B. larvae* from food by means of their proventriculus and/or the efficiency of a bactericidal factor in the gland secretions of nurse bees (Thompson and Rothenbuhler, 1957).

The most striking and unexpected fact about hygienic behaviour of the susceptible and resistant strains of bees selected by Rothenbuhler, was that the genes which determined both the prompt uncapping of cells and the efficient removal of the larvae in them were recessive. Hybrids of the two strains were all susceptible, and when these were cross-bred with the parent strains a quarter of the offspring were susceptible, a quarter were resistant, a quarter would uncap cells but did not remove the larvae, and a quarter would remove the larvae only when the cells were uncapped for them. This is explicable on the hypothesis that uncapping cells and removing larvae each depend for full expression upon homozygosity for two independent recessive genes. This unfortunately entails inbreeding, which, apart from its practical difficulties, can have undesirable consequences (Chapter 3, I.A.2.; Chapter 9, IV.).

The rate at which larvae of the resistant strains selected by Rothenbuhler became innately resistant to infection (Chapter 4, I.D.) was greater than that of other strains, possibly because they grew quicker during their first day of life (Sutter *et al.*, 1968) and because their food contained more factors that inhibited the germination of spores of *Bacillus larvae* and the growth of vegetative cells (Rose and Briggs, 1968a). Also the adult bees of the resistant strain filtered spores from the honey in their honey-sacs more efficiently than susceptible bees (Plurad and Hartman, 1965). Nothing is known of the genetic basis of these aspects of resistance, but Hoage and Peters (1969) were able to select for even greater resistance from the resistant strains by inoculating young larvae, especially drones, with a critical number, about 1000, of spores of *Bacillus larvae* and in-breeding from the survivors by artificial insemination.

The protective ability of adult bees either to filter off bacterial spores or to decrease their infectivity with bactericidal agents, or to do both, was also demonstrated in bees from Hawaii. American foulbrood was rampant in Hawaii during the early 1930s but was less evident in 1949 (Keck, 1949). It was thought that very resistant bees had survived the virtual abandonment of beekeeping in Hawaii during World War II, although 13 out of 86 colonies were found diseased when examined in 1949, which is still a high incidence. However, Eckert (1950) noted that the colonies stored a great deal of honey and that those found diseased before the nectar-flow became apparently healthy afterwards. Hundreds of colonies were then examined in abandoned apiaries, where combs had been largely destroyed by wax moths; and numerous swarms in cliffs and trees were also examined, but no disease was found. Resistance seemed likely, and Thompson and Rothenbuhler (1957) showed that bees from Hawaii protected larvae from American foulbrood about eight times more efficiently than colonies that had been selected for susceptibility were able to protect sister larvae. The diminished population of colonies in Hawaii after World War II probably decreased the rate at which American foulbrood had spread previously, and they probably fared much better in good years than they did before, thus being more able to ward off disease (Chapter 4, I.E.), but the increased hereditary ability of their bees to protect larvae may have been an important factor contributing towards their resistance.

There is no doubt that there are several inheritable factors that make some strains of bees more resistant than others to American foulbrood. The differences found between resistant bees and the average strains of bee are not sufficient to sustain hopes of eventually being able to select immune strains; and the task of separating the desirable from the undesirable characters, combining them and maintaining them, would be difficult in present circumstances. Nevertheless, the work of Rothenbuhler provides the best scientific evidence of hereditable resistance to disease in insects and is a solid foundation of knowledge about disease resistance in bees.

6. *The effects of beekeeping*

Infection by *Bacillus larvae* is the most unstable of all infections in honey bee communities. The balance is easily tipped in favour of its spread and it can be expected to increase once established in beekeepers' colonies, which have less opportunity to express the little resistance they have than undisturbed colonies. In view of the longevity of the spores of *B. larvae*, untreated equipment from diseased colonies should always be considered infective.

B. European Foulbrood

1. *Destruction of diseased colonies*

This has been required by law in some countries and, although it may help to decrease the incidence of the disease it has not been very effective and is almost certainly uneconomical. Destruction of colonies with either American or European foulbrood in Switzerland from 1908 to 1933 and in Britain from 1942 to 1967 did not decrease the incidence of European foulbrood, whereas the incidence of American foulbrood in Britain declined from about 7% of colonies to about 1% by 1954. (Min. of Agric., Fish. and Food, 1956). However, it is probably best to destroy severely diseased colonies, especially when the incidence of disease is low.

2. *Manipulative treatments*

Many have claimed that European foulbrood can be eradicated by removing the queens of diseased colonies for between 3 days and 3 weeks and then replacing them with newly mated queens (Langstroth, 1866; Phillips, 1921; Sturtevant, 1920). The idea is to give bees an opportunity to clear away diseased larvae and bacterial contamination. However, most diseased colonies recover anyway (Chapter 4, II.D.) and the queenless period almost certainly impairs the efficiency of this process. Dzierzon (1882) noticed that the larvae in the newly formed queen cells of dequeened colonies were frequently killed by European foulbrood, and queenless colonies keep more infected larvae than normal colonies (Chapter 4, IID). On the other hand, young queens tend to be prolific, which leads to the ejection of more infected larvae than usual by nurse bees. Requeening diseased colonies helps the process of spontaneous recovery but a queenless period is almost certainly valueless, and may be harmful.

3. *Sterilization of combs and honey*

Fumes of formaldehyde or acetic acid, applied in the same way that is advised for the disinfection of comb contaminated with spores of *Nosema apis* (Section IV. A.1.), will kill the resting stages of *Streptococcus pluton* when these are not buried beneath organic matter. Sound, empty comb from diseased colonies are worth treating in this simple fashion. Comb containing dead brood should be burned.

Ethylene oxide, applied as for disinfecting comb from colonies with American foulbrood (Section II. A.3.) will probably be at least equally effective for European foulbrood, but, curiously enough, gamma irradiation with up to 0·8 M rad is ineffective (Pankiw *et al.*, 1970).

4. Chemotherapy

Several different antibiotics effectively suppress European foulbrood, but the disease frequently recurs in treated colonies especially during the following season. Although recovery from the disease is accelerated by antibiotics, infection and contamination by *Streptococcus pluton* may not decrease as much as in colonies that are allowed to recover spontaneously. Many infected larvae, that would usually be ejected by adult bees, survive when treated with antibiotics, but they still discharge many infective bacteria in their faeces when they pupate. For best effect, antibiotics should be administered before the time when European foulbrood is usually seen, as early in the season as is practicable. Then the multiplication and accumulation of *Streptococcus pluton* is forestalled, disease outbreaks are decreased in severity, or do not occur, and residual infection is readily controlled by the usual activities of the bees (Chapter 4, II.D.).

Based on the above principle, a system of antibiotic treatment has been approved and administered officially in Britain since 1967. The percentage of colonies found with European foulbrood has remained as small as when destruction was mandatory, and now many of them are saved and remain productive.

The method is to give 0·5 or 1 g, to small or large colonies respectively, of oxytetracycline ("Terramycin") dissolved in about 500 ml of concentrated sucrose syrup, by sprinkling the syrup either over the bee cluster in the hive in warm weather, or into unoccupied comb taken from the edge of the colony in cold weather. These combs are replaced with the syrup side outwards from the bees to prevent them from consuming it too quickly, and to allow it to become mixed gradually with other food that is in the colony or given subsequently. This is necessary because the antibiotic is toxic for bees and brood in the concentration described (Gochnauer, 1954). It is still effective but harmless when more dilute, e.g. 1 g in 5–10 litres syrup, but it is impracticable to have such large volumes administered under official supervision. The policy is to treat diseased colonies when they are discovered, unless they are too severely diseased to be likely to recover, when it is recommended that they are burned. They are treated again as early as is practicable the next spring together with adjoining ("contact") colonies. No further treatment is given unless they suffer further outbreaks of disease, which have so far proved infrequent. Oxytetracycline decomposes in honey, which is safe for human consumption 8 or more weeks after treatment (Landerkin and Katznelson, 1957).

Other antibiotics could be used effectively in the same way, but they should not be used, either because they could be harmful for anyone consuming honey contaminated with them, or because those that are harmless should be held in reserve in the event of resistance developing towards oxytetracycline.

Other methods of application, such as dusting dry powders containing antibiotics or feeding several small doses at frequent intervals have been advocated, but none have been shown to be more effective than the method described above.

5. The effects of beekeeping

Methods of beekeeping that suppress brood-rearing or maintain a small amount of brood in proportion to nurse bees, allow *S. pluton* to multiply because they cause a temporary excess of glandular food supplied by nurse bees to larvae. For this reason disease is frequently severe in colonies that are regularly used for producing queen-cells or royal jelly.

In localities with uninterrupted nectar-flows, where colonies can grow unhindered each year, infection often remains slight and disease inapparent. However, when colonies are moved their development is checked. European foulbrood often becomes apparent after they have settled down in their new site and have begun to grow again. The disease often breaks out in endemically infected colonies after they have been brought back from the pollination of orchards in the spring (Bailey, 1961).

III. FUNGI

A. Chalk-brood

Chalk-brood is greatly aggravated by practices that cause the loss of heat from bee colonies, especially in spring and early summer when the natural incidence of chalk-brood is highest. Procedures that cause the loss of heat or that do not allow bees to maintain adequate temperatures have been used successfully to exacerbate chalk-brood in colonies for experimental purposes. For example, placing many combs of sealed brood from slightly diseased colonies into weak colonies, or removing many of the adult bees of infected colonies, or giving them extra brood to rear, have effectively increased the amount of disease (De Jong, 1976). The most usual equivalent practices in beekeeping, that cause trouble, are the methods employed to prevent swarming, which involve forcing the colony to expand either by dividing it or by spreading it into a larger space (Cale *et al.*, 1975). The colonies of beekeepers who do this routinely, or too early in the year, frequently suffer from chalk-brood. This is resolved when infected colonies are left undisturbed or when swarm control methods are delayed as long as possible. Chalk-brood recurs in such colonies, because the spores of *Ascosphaera apis* remain infective for many years, but its severity is greatly diminished.

There are no other known means of controlling chalk-brood, but spores in the larval remains are more easily killed by ethylene oxide fumigation than the spores of *Bacillus larvae* (Section II. A.3.) (Gochnauer and Margetts, 1980).

B. Other Fungi

No successful treatments are known.

IV. PROTOZOA

A. *Nosema apis*

1. *Hygienic methods*

Although it is theoretically possible to remove infection by transferring colonies to uncontaminated combs (Chapter 6, I.C.) in early summer, the method is costly in labour at least, and there are many factors, some of which have already been discussed, which can cause re-contamination. It is probably economically more sound to avoid aggravating the disease, rather than try to eradicate it, particularly as endemic infection is so widespread and difficult to detect.

Combs that contain no honey or pollen can be disinfected with formalin, but when there is food in the combs formalin makes it extremely poisonous for bees: then the most convenient safe fumigant is acetic acid (Bailey, 1955b; Hirschfelder, 1957; Jordan, 1957; Lunder, 1957). Commercial grades of either fluid may be used; acetic acid is available in a concentration of 80% which stays liquid below 15°C, unlike 100% or ''glacial'' acid. The easiest effective method of using these fumigants is to intersperse absorbent materials between piles of hive bodies containing the combs, and to pour about 150 ml of the liquid on to the material between each box. The stacks should be left in a warm corner out of doors and protected from direct winds for a few days before being used. It is best to air them for a day before putting them in colonies.

Keeping contaminated combs at 49°C and 50% relative humidity for 24 h greatly decreases their infectivity without damaging them (Cantwell and Lehnert, 1968; Cantwell and Shimanuki, 1970).

2. *Chemotherapy*

More drugs have been tested against *Nosema apis* than against any other parasite of honey bees, largely because of the ease with which laboratory tests can be made. Almost all have proved ineffective. (Katznelson and Jamieson, 1952b, 1955; Bailey, 1961b, 1962; Palmer-Jones and Robinson, 1951; Moffatt *et al.*, 1969).

The most outstandingly successful drug so far is the antibiotic fumagillin, which is derived from *Aspergillus fumigatus* and prepared and marketed under the trade name ''Fumidil B''. It markedly decreases infection when fed continuously to caged bees in concentrations between 0·5 and 3 mg/100 ml syrup, with no ill effects. (Katznelson and Jamieson, 1952a). The activity of the antibiotic remains high in honey kept at 4°C for several years (Furgala and Gochnauer, 1969a). When about 200 mg of fumagillin in 4·5 or 9 litres of syrup is fed to each colony in autumn, infection is markedly decreased the next spring (Bailey, 1955c). When fed in spring, fumagillin prevents the usual peak of infection and treated colonies have produced significantly more brood and honey than similar colonies fed syrup only (Farrar, 1954; Gochnauer, 1953; Jamieson, 1955; Furgala and Gochnauer, 1969b). Spring feeding may not have such a long-term effect as autumn feeding because spores, which are mostly deposited on combs in winter (Chapter 6, I.C.), are unaffected by the antibiotic. Applying the antibiotic as a dust or in solid sugar candy is ineffective (Furgala and Gochnauer, 1969b). Attempts to select strains of *N. apis* resistant to fumagillin, by maintaining infection in bees treated with subliminal doses of the antibiotic, have failed (Gross and Ruttner, 1970).

A variety of mercurial compounds of the type sodium ethyl-mercurithiosalicylate, commonly known as "Merthiolate", have been used successfully to suppress infection when fed in autumn or spring (Gontarski, 1954). Laboratory experiments indicate, however, that although these compounds are effective against *Nosema apis* they are poisonous for bees (Bailey, 1961b). Moreover, they would be dangerous contaminants of honey for human consumption.

Other drugs reported to have a good effect are: gramicidin (Alpatov and Kiriakova, 1953); sulphaquinoxaline and anisomycin (Katznelson and Jamieson, 1955); phenol uratropin and β-naphthol (Alpatov, 1942); β-naphthol and salol (White, 1919); and colloidal sulphur (Falkener, 1939). All need confirmation, and their toxicities need to be checked, even if they are effective against *Nosema apis*.

3. The effect of beekeeping methods on infection

There are several ways in which beekeeping practices aggravate infection. Contaminated combs placed in colonies towards the end of summer, as they often are during late nectar-flows, will reintroduce infection (Fig. 16), too late for the bees to clear it up adequately by autumn. Severely infected clusters will then dwindle more rapidly than usual during the winter or may even die.

Individual bees are frequently crushed when colonies are opened and examined, and they are then removed by the other bees which ingest the liquid remains. This aggravates any infection and probably explains the high incidence of *Nosema apis* in "package bees" (Chapter 6, I.D.). These are small clusters or colonies of bees produced in warm regions, such as southern U.S.A., for sale to beekeepers in northern areas where it is more expensive to keep colonies throughout the winter than to destroy them in autumn, take all the honey and buy more bees the following spring. The constant handling suffered by the parent colonies, followed by the confinement of the bees in the package, maintain severe infections (Farrar, 1947; Reinhardt, 1942).

Infection is aggravated by transporting colonies to new sites where there may be desirable nectar-flows or where pollination is required (Bailey, 1955b). The bees often end such journeys in a distressed condition, and many then deposit faecal matter within the colony, even on the cappings of newly sealed honey.

The most dangerous time to disturb colonies is during late winter and early spring when the bees have been confined for long periods (Table VIII). In practice, infection is most likely to be spread by moving or disturbing colonies during the spring, because spontaneous infection of bees has become most severe at this time when they are most actively cleaning contamination from the combs. The practice of taking colonies to fruit orchards for spring pollination has considerably increased in recent years and probably causes the severe infections suffered by colonies of many beekeepers engaged in this work. Many beekeepers also frequently examine their colonies too soon in spring and early summer.

Commercial beekeepers who are compelled to move and handle colonies a great deal may find it advantageous to practise a rotation system, keeping some hives undisturbed at good sites for a year and allowing the natural control of infection to take full effect.

Queen rearing is another beekeeping practice likely to aggravate infection. The small colonies of bees used to keep and mate virgin queens (''mating nuclei'') are populated by bees that are older than average and so more likely to be infected. This is because little brood is produced by the newly mated queens before they are replaced with another to be mated, and the colonies are also frequently handled (Table VIII).

TABLE VIII. The number of bee colonies found infected with *Nosema apis* at the end of April in two successive years at Rothamsted.

Group	Treatment	Year			
		1		2	
		Infected	Healthy	Infected*	Healthy
1	Colonies transferred to disinfected combs during early summer of Year 1 and increased by division	12	0	5	18
2	Combs examined monthly for other tests during winter of Year 1/Year 2	3	11	10	0
3	Colonies without queens for several weeks during summer of Year 1	4	11	8	2
4	None	79	111	71	109

* The proportion of infected colonies in Year 2 changed significantly from that in Year 1 in all groups except 4, showing that: (Group 1) transferring colonies to disinfected combs removes much residual infection; (Group 2) disturbing colonies aggravates infection and (Group 3) colonies without queens become more infected than usual (original data).

The same effects caused by poor seasons on infection are brought about by keeping too many colonies together, which then cannot support themselves on the available forage. This may largely explain the high proportion of infection in urban regions, for instance the 30% of colonies in the urban south-east of England compared with 3–16% in the other areas of England and Wales (Min. of Agric., Fish. and Food 1956, 1957, 1958, 1959), and 30–90% of colonies infected in experimental stations in England, Scotland and Russia (Morison *et al.*, 1956). These high incidences of infection are also aggravated by the unusual amount of handling suffered by many of the colonies.

B. *Malpighamoeba mellificae*

1. *Hygienic methods*

Infection can be removed by transferring colonies in early summer to non-contaminated combs (Fig. 17), a procedure less likely to fail with *M. mellificae* than with *Nosema apis* (Chapter 6, II.C). However, it is probably more economical to suppress rather than to try to eliminate infection. This can be done by fumigating spare combs in the same way as for *Nosema apis* (Section, IV.A.1.).

2. *Chemotherapy*

Few attempts at chemotherapy have been made. Field experiments have shown the antibiotic fumagillin to be ineffective (Bailey, 1955d). Salol, quinosol, furazlidone and dichloroxyquinaldine have also proved ineffective (Giordani, 1959).

3. *The effect of beekeeping methods on infection*

The equivalent section for *Nosema apis* (Section IV. A.B.) applies equally to *M. mellificae.* In Britain, most infection (3% of colonies) is in the south-east near London, probably for the same reasons as for *Nosema apis*. In Denmark too, most infection by *M. mellificae* has been found near towns, particularly near Copenhagen (Fredskild, 1955). Nevertheless, infection by *M. mellificae* remains low compared with that by *Nosema apis*, probably because cysts develop slowly, maturing only when bees are nearing the end of their normal summer lives, so there is far less likelihood of combs becoming contaminated with them in summer than with spores of *N. apis*. (Chapter 6, II.B.).

V. PARASITIC MITES

A. *Acarapis woodi*

1. *Hygienic methods*

As brood is not infested it may be separated, when sealed, from infested colonies and used to create new uninfested ones by hatching it in incubators and providing a new uninfested queen; or it may be added to uninfested colonies. This idea, with added complications believed to increase efficiency, has been advocated many times. The principle has been tested on a large scale in Europe and has apparently completely eradicated the mites (Kaeser, 1952). The method is laborious, however, and involves the loss of bees and queens.

2. *Chemotherapy*

The fumes of burning sulphur, produced by smouldering sulphur-impregnated fuel in a bee smoker, was recommended by Rennie (1923) and was found at least partially successful by some (Frow, 1927; Morganthaler, 1948; Ingold, 1950; Gubler *et al.*, 1953), but not by others (Sanctuary, 1924; Lavie, 1950; Kaeser, 1952). It probably is effective, but in circumstances that still need to be carefully determined. The fuel is prepared by cutting corrugated brown paper strips about 3 in wide, across the corrugations. These are soaked in 30% aqueous solution of potassium nitrate and dried. A thick smooth paste of flowers of sulphur mixed in a hot solution of glue-size (60 g of size to 500 ml of water) is brushed on the paper strips which are then partially dried and made into rolls that will fit into a bee smoker. Three puffs of smoke given each day for 10 successive days, followed by a similar treatment after a 7-day interval, have given good results, and shorter treatments may also be effective (Ingold, 1950).

The vapour of a mixture of petrol, nitrobenzene, and safrol, in the ratio by volume of 2:2:1 respectively, "Frow's mixture" (Frow, 1927), can kill mites without causing obvious harm to bees. About 2 ml of the liquid is poured on an absorbent pad over the centre of the cluster of bees in early or late winter when they are more or less inactive. About 7 treatments at 2-day intervals are necessary. The treatment is dangerous when bees are active, because it kills unsealed brood and because the treated colonies become attractive to robber bees. However, the vapour is not harmless to adult bees, even when they are inactive. In controlled laboratory tests more than half the treated bees were killed by the concentration of vapour found necessary to kill all mites in the survivors (Morganthaler, 1931b). In the short time taken by these tests, the proportion of infested bees in the survivors was the same as before, so the vapour seemed not to kill infested bees any more easily than those that were uninfested. However, field experiments have shown that infested bees begin to die within a month of treatment applied in early winter, and all such bees have died by the beginning of the new year, which is much earlier than usual (Bailey and Carlisle, 1956). Persistent annual treatment of this kind should eliminate infestation, but infested colonies are more likely to die after treatment than they would without it and it is questionable whether this loss would be recouped by the increased performance of the survivors.

Many research workers on the European continent have reported Frow's mixture to be unsatisfactory, so environmental conditions may have much to do with its effectiveness. The best conditions are probably those when the vapour percolates the cluster slowly in concentrations strong enough to kill mites but not to disturb the bees unduly. Autumn or spring weather in Britain may be more suitable than it is elsewhere in this respect.

The vapour of methylsalicylate has been found effective against mites. The vapour damages unsealed brood, however, and although controlled laboratory tests have shown that it can kill significant numbers of mites, the concentration needed is not far below that which is obviously toxic for bees (Bailey and Carlisle, 1956).

Specific acaricides, with no apparent side-effects on larval or adult bees, are now available. The best known are "Dimite" ((1,1 bis (p–chlorphenyl) ethanol) and "Chlorobenzilat" ((di–p–chlorophenyl)-glycollic acid ethyl ester), prepared for use with bees under the trade names "P.K." and "Folbex". Strips of paper about 11 × 4 cm are impregnated with potassium nitrate and about 500 mg of the acaricide. The end of a paper is ignited, and it then smoulders, giving off a smoke containing the volatilized acaricide. It is hung inside the hive, burning edge down, and the entrance of the hive is closed for at least half an hour. These materials are best used in warm weather, when one or two treatments will suppress infestation and sometimes almost eliminate it (Bailey and Carlisle, 1956).

Much effort has been expended trying to find suitable fumigants, and eradication rather than suppression of infestation has been considered the only desirable objective. The economy of this aim is questionable. Severe infestations build up only in colonies that, for one reason or another, are not doing particularly well and would not be profitable even when uninfested. Where widespread and severe infestation occurs regularly, it is probably unprofitable to keep bees at all. There may be marginal regions where it is worth keeping bees but where infestation occasionally develops enough to affect the economy. Even so, it is probably unnecessarily expensive to try to eliminate the mite which would almost certainly re-invade the cleared area.

3. *Resistant strains of bees*

Almost certainly there are genetic differences between strains of bees which cause differences in their susceptibility to infestation; but a comparison of bees of an allegedly resistant strain with local hybrids showed no significant difference between them (Bailey, 1961a). Any differences are probably slight and may require several seasons to show a cumulative effect.

4. *The effect of beekeeping methods on infestation*

Infestation is aggravated by procedures that check the growth and foraging of colonies. Infested colonies that suffer periods of queenlessness in early summer become more severely infested than colonies with queens (Bailey and Lee, 1959). The numbers of the external mite *Acarapis vagans* also increase in queenless colonies (Schneider, 1941). The numbers of the other *Acarapis* species probably increase similarly (Chapter 7, III.).

Keeping too many colonies for the available forage, and the frequent closure of hives may be expected to increase infestation by increasing contact between the older, more infested bees, which would normally be out foraging, and young susceptible individuals.

B. *Varroa jacobsoni*

Many different chemical treatments, some of which have been used by beekeepers for several years, have been tested and compared by Ruttner and Ritter (1980). They tried a wide variety of compounds applied as a vapour, in smoke, in sprays and as systemic acaricides. None was entirely satisfactory and many were harmful to the bees. Any that were effective were least efficient when brood was present, because many mites were then protected within the sealed cells. The most promising material seems to be chlordimeform hydrochloride, a systemic acaricide applied as a solution of 35 mg in 50 ml water, dripped or sprayed on to the bee cluster in the autumn immediately after they have been fed for the winter and when they have little or no brood. Damage to adult bees seems insignificant.

More than one application may be necessary. The acaricide is quickly degraded and, although this is desirable from the point of view of not having residues that might contaminate honey, it does not persist long enough to protect larvae that are sealed in their cells with mites. A similar but more persistent systemic acaricide would seem to be the ideal. Clinch (1979) and Clinch and Faulke (1977) tested several against external species of *Acarapis* (Chapter 7, III.) and found endosulfan fed in syrup much more effective than chlordimeform, although endosulfan would be an unacceptably toxic contaminant of honey for human consumption.

Ruttner and Ritter (1980) advocate a method of decreasing the amount of brood in infested colonies to prevent migrating females of *V. jacobsoni* from finding protection in sealed cells against chemical treatments. The method is to confine the queen on one or two empty brood-combs, preferably of drone cells, surrounded by combs full of sealed stores or by frames of wax foundation on which the queen does not lay. The queen can be prevented from returning to the main mass of brood by means of wire grids ("queen-excluders") with slots wide enough for workers but not queens to pass through. The combs on which she lays are removed, when the brood in them has been capped, and are replaced with empty ones. The process is continued for about a month by the end of which the colony has no brood except in the combs on which the queen is confined. These trap most of the migrating mites that have escaped chemical treatment and they are removed and destroyed with the combs. The method is ingenious, but it destroys a month's production of brood. This, the labour involved, and the aggravating effect the method would have on other infections, such as those by *Nosema apis* and *Acarapis woodi*, are severe disadvantages.

VI. INSECT PESTS

A. Wax Moths

One of the most efficient fumigants against wax moths is ethylene dibromide (Krebs, 1957). Boxes of comb should be enclosed in plastic bags into which the liquid ethylene dibromide is added. Between 20 and 100 g of ethylene dibromide per 1000 litres of volume to be fumigated is recommended. The liquid, which is poisonous to man and a non-inflammable vesicant, should be poured in at the top of the pile. Methyl bromide is very effective too (Roberts and Smaellie, 1958), but is dangerously poisonous to man and is a gas at normal temperatures, which makes it less convenient than ethylene dibromide.

Paradichlorbenzene is safe for man and its crystals give off vapour which quickly kills both adults and larvae of wax moths at ordinary temperatures. Eggs are much more resistant to it, however, and combs are repellent to bees for a short time after treatment.

Acetic acid, as used for fumigating combs against contamination by *Nosema apis* or *Malpighamoeba mellificae* (Section IVA1) quickly kills eggs and adults of wax moths. However, larvae especially the largest of *Galleria mellonella*, are resistant, particularly when they are buried in comb debris.

Heating combs at 49°C for 24 h at 50 % relative humidity, as suggested for disinfecting them of *Nosema apis* spores (Section IV A1), kills all stages of wax moths (Cantwell and Smith, 1970).

Cooling combs to temperatures between 0°C and – 17°C, for a few hours to several days, according to temperature and bulk of material, kills all stages of wax moths without harming combs or stored honey (Burges, 1978).

Preparations of *Bacillus thuringiensis* are effective against larvae of wax moths, and can be incorporated into the wax foundation of combs, but the persistence of the bacilli is insufficient to make economic the use of currently available strains (Burges, 1978).

B. Other Insects

According to Boiko (1949) adult flies that give rise to apimyiasis rest on hive roofs, especially when these are light in colour, whereas bees do not, and a recommended control measure is to paint the roofs with a persistent insecticide.

Naphthalene sprinkled on paper on the hive floor is said to kill oil beetle larvae without harming bees (Minkov and Moiseev, 1953), and a long established practice said to relieve queen bees of infestation by *Braula coeca* is to expose them briefly to tobacco smoke.

VII. INSECTICIDES

The only satisfactory way to avoid losses caused by insecticides is to keep bee colonies a mile or more away from areas that are to be sprayed, and to spray crops when bees are not attracted to them. These are matters requiring close and friendly relationships between beekeepers and farmers rather than legislation. When colonies cannot be moved from an area to be sprayed they can be protected by covering them, including the hive entrance, during daylight hours, with coarse sacking material. The sacking should be kept soaked with water to keep the colonies as cool as possible (Wilson *et al.*, 1980).

11. CONCLUSIONS

A wide variety of specific pathogens are endemic in honey bees, and most of them are perpetuated as inapparent infections. Colonies, and even individual insects, that are infected with certain, sometimes several, pathogens frequently seem outwardly normal for an indefinite period. Accordingly, it is often difficult to identify the causes of losses and disorders, and this has led to much confusion and many false diagnoses; especially when some pathogens, particularly viruses, have gone unrecognized. Understandably, severe disorders of colonies are the first to preoccupy attention. They were the ones that were first investigated when the study of honey bee pathology was begun seriously in the late nineteenth and early twentieth century, and they gave rise to the belief that pathogens of bees always cause diseases as severe as those that had been recently diagnosed in silkworms and other domesticated animals. In fact, apparently healthy colonies of bees resemble all other populations of wild animals, including pest insects, by sustaining many different endemic infections and in most circumstances by not becoming crippled by them.

Nevertheless, the sum of damage, including premature changes to the behaviour of individuals, brought about by the widespread, usually inapparent or unnoticed infections of bees, probably exceeds the loss of the comparatively few colonies that become severely diseased or die. The many and inevitable losses of worn-out adult individuals by all bee colonies and their rapid replacement in growing colonies during the active season obscure the additional, sometimes severe, losses caused by disease.

The available evidence shows that honey bee pathogens multiply and spread to cause more damage than usual within those colonies that are most hindered in their development by adverse environmental events. Food shortage is the first result of these unfavourable conditions for bees in nature, and it is often associated with an abnormal increase of pathogens. This is analogous to the nature of infectious diseases in other wild animal communities, in which severe disease is generally associated with food shortage and is a secondary cause of death (Lack, 1954a, 1954b).

In nature, endemic infections of these kinds may be a net benefit to the species by decreasing the chances for survival of strains that are least well adapted to the environment. This is equivalent to the favourable selective pressure that predators exert on wildlife by attacking the least vigorous prey. However, the balance that has been achieved by even the best-adapted strains of bees with their pathogens is an uneasy one and is readily deranged, especially by beekeeping practices. These can impede or interrupt the development of colonies much more than natural events and quickly lead to severe losses from disease. It is important to realize that the pathogens do not usually initiate the setbacks. Although pathogens, by definition, are the essential causes of infectious diseases, they are usually unable to multiply sufficiently to cause severe disease without supplementary factors. When they have the opportunity to multiply and spread, they will overwhelm a colony; but they are usually prevented by several mechanisms from doing this in colonies that are otherwise normal and in good circumstances.

There is no evidence that insects, including bees, are protected individually by a well-developed ability to acquire immunity, as are vertebrate animals. The spread of infections between bees is usually limited by an innate immunity of individuals during much of their life coupled with events that decrease the chance of contact between healthy individuals and the pathogens, such as the separation of old individuals, which are most likely to be infected, from young susceptible ones; and by their short life and expendable nature. In bee colonies, the oldest individuals are foragers, which become separated to a large extent from young bees during good foraging conditions. Moreover, diseased individuals are ejected by their companions or die away from the colony and the pathogens in them then have little chance of reinfecting others.

Colonies in natural habitats are probably more widely dispersed on the whole than when kept by man, and so they are less likely to suffer from a shortage of nectar and pollen, and the consequent inactivity that in turn leads to the spread of pathogens within them. Furthermore, nests of wild colonies that become weak and die are less likely than beekeepers' colonies to be found and ransacked or occupied by other bees before they are found and destroyed by scavengers. Such events again decrease the chance of contact between healthy bees and pathogens. Indeed the first principle in the control of all diseases is to minimize contact between the host and the pathogens. It is unnecessary, often impossible, for them to be separated absolutely, but when the balance between the spread of a pathogen and the forces that oppose it is tipped sufficiently in favour of the host, the natural system will itself do the rest and disease will be controlled, or even completely suppressed. Pasteur was able to eradicate pebrine from the silkworm because he could select healthy stocks of insects and then keep them easily and permanently in isolation. This can rarely be done, and then often only with difficulty, even with domesticated animals. It cannot be done with feral animals such as bees, as has been explained previously (Chapter 1); so it is unreasonable to expect to eradicate their diseases easily. Nevertheless a great deal can be done to avoid aggravating them and to assist the natural mechanisms that oppose their spread.

A common feature of several different infections of bees is that the pathogens are transmitted between seasons as resting stages on the comb. In nature, these are cleaned by adult bees or destroyed by scavengers, and infection is thereby usually controlled. However, beekeepers began seriously to interfere with this process when they exchanged their hives with fixed combs, which had to be destroyed when their honey was extracted, for those with moveable combs, which are durable. These combs are frequently stored away from bees, to be returned to them when they need more space. This interrupts the natural cleaning process and reinoculates pathogens into the colonies when the combs are returned. Often this occurs too late in the season for the bees to recover from the infection as efficiently as usual. The efficient sterilization of combs or even the destruction of those known to be severely contaminated is the obvious solution.

The popular preference, or hope, has always been to treat disease with drugs; and the advent of effective specific ones, particularly the sulphonamides and antibiotics that act so dramatically against bacteria, seems to have justified this desire for spectacular, almost magical results. Various nostrums have always been recommended, long before the advent of effective antibiotics, even by Virgil (70–19 B.C.) who advised that various herbs and wine should be fed in honey to ailing colonies; but there is no evidence that any are of benefit. Like so many alleged remedies, their apparent effectiveness most probably depends upon the innate power of animals to recover from disease, sometimes in spite of treatment. Indeed, the damage to bees caused by many treatments has often been believed to be the result of infections too advanced to be cured and has increased the notoriety of bee diseases. This applied particularly to *Acarapis woodi* during the early part of the century and probably had much to do with the alarm about *Varroa jacobsoni* more recently.

Even effective antibiotics can be toxic to bees and are often seriously misused. Furthermore, it can be pointless, or at least inadvisable, to use them against diseases that usually decline spontaneously when they are known to have reached their seasonal peak. Antibiotics are then least likely to affect the course of infection or to prevent contamination of comb by dormant organisms, and they are most likely to contaminate honey harvested for human consumption. Their use against severe, or potentially severe, diseases such as American or European foulbrood is inadvisable when the incidence of disease is very low. It is then more economical to destroy the few colonies that become diseased. Otherwise, the use of antibiotics, which do not eradicate infection, allows pathogens to spread undetected and leads to dependence on frequent and widespread treatment.

Virus diseases seem to pose a less tractable problem than the rest. Viruses of bees do not appear to have resting stages. They do not survive long outside the living tissue in which they multiply, and, in common with all viruses, they are not susceptible to treatment with available antibiotics. Knowledge of their ecology is scanty, but there are indications that their spread between individuals within colonies is increased by unfavourable environmental conditions, especially those that keep bees confined to their cluster when they normally would be flying freely. Some are secondary infections, dependent upon primary infection by *Nosema apis*, which makes the control of this parasite a more important objective than it once seemed to be. Although no direct therapy is yet known for virus infections it is necessary to be aware of them if only to avoid mistaken diagnoses and to avoid the importation of exotic types and strains. Further knowledge of them is needed, and their successful management could well be the reward of a better understanding of their ecology.

All diseases are influenced by genetic factors, but there is no evidence that any strains of honey bees have genes that give them immunity from any of their known pathogens. Beekeepers have occasionally claimed that their bees were immune because the bees did not succumb, or because they recovered quickly after being exposed to certain diseases. The beekeepers were impressed, because they have been led to believe that bees are usually very susceptible, whereas their observations confirmed that bees have considerable powers of resistance and recovery. Even were immune bees to exist, it would be very difficult to replace common strains with them before virulent mutants of the pathogens found their way back from the reservoir of susceptible bees. The genetic variability of pathogens and the possibility of the inadvertent selection of more virulent ones is frequently overlooked. Combs contaminated with pathogens, possibly of greater virulence than usual, are frequently removed from dead colonies and used again instead of being destroyed, as many would be in nature by scavengers. Colonies that are ailing when others are not, and that are often helped to survive by beekeepers, may also be infected with unusually virulent pathogens which would be better destroyed.

There is no doubt that bees can be selected with resistance that is greater than average towards disease, as has been demonstrated by Rothenbuhler and his colleagues with American foulbrood. It seems significant, however, that resistance towards this disease, which is the most likely of all diseases to kill a bee colony, is at least partly determined by several recessive factors (Chapter 10, II.A.5.). This suggests that death of infected colonies has enabled the species as a whole to survive better than has natural selection for resistance. Whatever the reason, it makes the maintenance of resistant strains more difficult than if resistance were due to dominant genes. Attempts to find strains of bees more resistant to disease will inevitably continue, but the cost of maintaining and propagating them has to be weighed against the cost of the disease. Susceptibility to a disease that is slow to spread between colonies, is of low incidence and is easy to see could well be preferable to resistance that allowed it to become widely distributed in a form often difficult to detect.

It is a common belief that pathogens eventually become avirulent, or nearly so, after long association with a host, and some honey bee pathogens have been thought to provide examples. However, a pathogen will evolve in the way that best secures its chances of survival. Virulence may be essential for it to achieve this. For example, *Bacillus larvae* must kill the individual it infects in order to form durable and infective spores. In nature, before beekeeping began to influence its distribution, its spread between colonies must have been very slow, otherwise all bees would have been destroyed. The bacillus may originally have been localized to cool habitats, where wax moths do not destroy combs as quickly as they do in warm climates. At the other end of the scale, *Acarapis woodi* is only slightly harmful to adult bees. Its life cycle is long, relative to that of its host which has to survive in order to transmit migrating mites to other bees. Greater virulence would be disadvantageous for the mite, although it remains sufficiently harmful to damage or even kill the relatively few colonies that become severely infested.

The artificialities of beekeeping will no doubt increase as agriculture develops, particularly as it grows in scale and gives extensive areas of single, nectar-yielding crops for limited periods. This encourages beekeepers to move their bees from one crop to another, so inhibiting the development and normal activities of colonies, and to keep them too crowded in apiaries, especially between the periods of major nectar-flows. Such practices encourage the spread of infections within and between bee colonies. Furthermore, the common wish of beekeepers to import bees that are alleged to be superior to their own is easily gratified with the aid of modern transport, but little heed has been paid to the dangers of introducing exotic diseases or unusual strains of pathogens, especially of viruses, which are not easily diagnosed. For these reasons more attention needs to be paid than in the past to the prevention and suppression of diseases. Only by understanding the natural histories of the pathogens will effective or improved methods be devised. The difficulties to be overcome may be great, but so is the room for improvement.

REFERENCES

Alfonsus, E.C. (1935). *J. Econ. Entomol.* **28**, 568–576.
Alpatov, V.V. (1942). *Zool. Zh.* **21**, 147–152.
Alpatov, V.V. and Kiriakova, V.A. (1953). *Pchelovodstvo,* **5**, 41–44.
Amici, A.M. and Vecchi, M.A. (1958). *Ann. Microbiol. Enzimol.* **8**, 93–97.
Anon. (1981) *Glean. Bee Cult.* **109**, 156 only.
Aristotle. 384–322 B.C.). "History of Animals" Book IX, 262. Translated by Richard Crosswell, (1907). George Bell, London.
Atwal, A.S. (1967). Tech. Rep. Coll. Agr. Punjab. Agric. Univ. Ludhiana, 39 pp.
Bailey, L. (1954). *J. Exp. Biol.* **31**, 589–593.
Bailey, L. (1955a). *Parasitology,* **45**, 86–94.
Bailey, L. (1955b). *Ann. Appl. Biol.* **43**, 379–389.
Bailey, L. (1955c). *Bee World,* **36**, 121–125.
Bailey, L. (1955d). *Bee World,* **36**, 162–163.
Bailey, L. (1958). *Parasitology,* **48**, 493–506.
Bailey, L. (1959a). *Bee World,* **40,** 66–70.
Bailey, L. (1959b). *J. Insect Pathol.* **1**, 80–85.
Bailey, L. (1959c). *J. Insect Pathol.* **1**, 347–350.
Bailey, L. (1960). *J. Insect Pathol.* **2**, 67–83.
Bailey, L. (1961a). *Bee World,* **42**, 96–100.
Bailey, L. (1961b). Rep. Rothamsted Exp. Sta. 1960, 175 only.
Bailey, L. (1962). *Rep. Rothamsted Exp. Sta. 1961*, 160–161.
Bailey, L. (1963a). "Infectious Diseases of the Honey-bee". Land Books, London.
Bailey, L. (1963b). *J. Gen. Microbiol.* **31**, 147–150.
Bailey, L. (1965a). *J. Invertebr. Pathol.* **7**, 141–143.
Bailey, L. (1965b). *J. Invertebr. Pathol.* **7**, 167–169.
Bailey, L. (1966). *J. Apicult. Res.* **5**, 127–136.
Bailey, L. (1967a). Rep. Rothamsted Exp. Sta. 1966, 215 only.
Bailey, L. (1967b). *J. Apicult. Res.* **6**, 99–103.
Bailey, L. (1967c). *In* "Proceedings of the International Colloquium on Insect Pathology and Microbial Control" (P.A. van der Laan, ed.) 162–167. North Holland, Amsterdam.
Bailey, L. (1967d). *Ann. Appl. Biol.* **60**, 43–48.
Bailey, L. (1967e). *J. Apicult. Res.* **6**, 121–125.
Bailey, L. (1968). *J. Invertebr. Pathol.* **12**, 175–179.
Bailey, L. (1969a). *Ann. Appl. Biol.* **63**, 483–491.
Bailey, L. (1969b). *Bee World,* **50**, 66–68.
Bailey, L. (1972). *J. Apicult. Res.* **11**, 171–174.
Bailey, L. (1974). *J. Invertebr. Pathol.* **23**, 246–247.
Bailey, L. (1975). *Bee World,* **56**, 55–64.

Bailey, L. (1976). *Advan. Virus Res.* **20**, 271–304.

Bailey, L. (1977). Rep. Rothamsted Exp. Sta. 1976, 104 only.

Bailey, L. (1981). Unpublished information.

Bailey, L. and Ball, B.V. (1978). *J. Invertebr. Pathol.* **31**, 368–371.

Bailey, L. and Carlisle, E. (1956). *Bee World*, **37**, 85–94.

Bailey, L. and Collins, M.D. (1981). In Press.

Bailey, L. and Fernando, E.F.W. (1972). *Ann. Appl. Biol.* **72**, 27–35.

Bailey, L. and Gibbs, A.J. (1962). *J. Gen. Microbiol.* **28**, 385–391.

Bailey, L. and Lee, D.C. (1959). *J. Insect Pathol.* **1**, 15–24.

Báiley, L. and Woods, R.D. (1974). *J. Gen. Virol.* **25**, 175–186.

Bailey, L. and Woods, R.D. (1977). *J. Gen. Virol.* **37**, 175–182.

Bailey, L., Gibbs, A.J. and Woods, R.D. (1963). *Virology*, **21**, 390–395.

Bailey, L., Fernando, E.F.W. and Stanley, B.H. (1973). *J. Invertebr. Pathol.* **22**, 450–453.

Bailey, L., Ball, B.V. and Woods, R.D. (1976). *J. Gen. Virol.* **31**, 459–461.

Bailey, L., Carpenter, J.M. and Woods, R.D. (1979). *J. Gen. Virol.* **43**, 641–647.

Bailey, L., Ball, B.V., Carpenter, J.M. and Woods, R.D. (1980a). *J. Gen. Virol.* **46**, 149–155.

Bailey, L., Ball, B.V. and Perry, J.N. (1980b). *Ann. Appl. Biol.* **97**, 109–118.

Bailey, L., Carpenter, J.M., Govier, D.A. and Woods, R.D. (1980c). *J. Gen. Virol.* **51**, 405–407.

Bailey, L., Carpenter, J.M. and Woods, R.D. (1981). *J. Invertebr. Pathol.* In Press.

Bamrick, J. (1967) *J. Invertebr. Pathol.* **9**, 30–34.

Bamrick, J.F. and Rothenbuhler, W.C. (1961). *J. Insect Pathol.* **3**, 381–390.

Barker, R.J. (1978). *In* "Honey Bee Pests Predators and Diseases" (R.A Morse, ed.) 276–296. Comstock, Ithaca.

Batuev, Y.M. (1979). *Pchelovodstvo*, **7**, 10–11.

Beljavsky, A.G. (1933). *Bee World*, **14**, 31–33.

Benoit, P.L.G. (1959). *Bee World*, **40**, 156 only.

Betts, A.D. (1912). *Ann. Bot. (London)*. **26**, 795–799.

Betts, A.D. (1932). *Bee World*, **1**, 132 only.

Beutler, R. and Opfinger, E. (1949). *Z. Vergl. Physiol.* **32**, 383–421.

Blum, M.S., Novak, A.F. and Taber, S. (1959). *Science*, **130**, 452–453.

Boiko, A.K. (1948). *Dokl. Akad. Nauk. SSSR*. **61**, 423–424.

Boiko, A.K. (1958). Int. Congr. Beekeep. Proc. 17th, 24–25.

Borchert, A. (1929). *Bee World*, **10**, 149 only.

Borchert, A. (1948). *Deut. Bienen-Ztg*. **3**, 113–114.

Braun, W. (1957). *Apic. Am.* **2**, 9–14.

Breed, R.S., Murray, E.G.D. and Smith, N.R. (1957). "Bergey's Manual of Determinative Bacteriology". 7th Ed. Balliere, Tindall and Cox, London.

Brügger, A. (1936). *Arch. Bienenkunde*, **17**, 113–142.

Buchanan, R.E. and Gibbons, N.E. (1974). "Bergey's Manual of Determinative Bacteriology" 8th Ed. Williams and Wilkins, Baltimore.

Bulger, J.W. (1928). *J. Econ. Entomol.* **21**, 376–379.

Bulla, L.A. and Cheng, T.C. (1976). "Comparative Pathobiology. Volume l. Biology of the Microsporidia". Plenum Press, New York.

Bullamore, G.W. (1922). *Parasitology*, **14**, 53–62.

Bullamore, G.W. and Maldon, W. (1912). *J. Board. Agric. G.B.* **19**, supp. 8, sects. 2 and 3.

Burges, H.D. (1978). *Bee World*, **59**, 129–138.

Burnside, C.E. (1928). *J. Econ. Entomol.* **21**, 379–386.

Burnside, C.E. (1930). U.S. Dep. Agr. Tech. Bull. No. 149

Burnside, C.E. and Vansell, G.H. (1936). U.S. Bur. Entomol. Rep. No. 398.

Butler, C.G. (1943). *Bee World*, **24**, 3–7.
Buys, B. (1972). *South African Bee Journal*, **44**, 2–4.
Cale, G.H., Banker, R. and Powers, J. (1975). *In* "The Hive and the Honey Bee" (Dadant and Sons, eds). 355–412, Dadant and Sons, Hamilton.
Cantwell, G.E. and Lehnert, T. (1968). *Am. Bee. J.* **108**, 56–57.
Cantwell, G.E. and Shimanuki, H. (1970). *Am. Bee. J.* **110**, 263 only.
Cantwell, G.E. and Smith, L.J. (1970). *Am. Bee J.* **110**, 141 only.
Cheshire, F.R. and Cheyne, W.W. (1885). *J. Roy. Microsc. Soc.* **5**, 581–601.
Clark, T.B. (1977). *J. Invertebr. Pathol.* **29**, 112–113.
Clark, T.B. (1978). *J. Invertebr. Pathol.* **32**, 332–340.
Clinch, P.G. (1974). *N.Z.J. Exp. Agr.* **2**, 451–453.
Clinch, P.G. (1976). *N.Z.J. Exp. Agr.* **4**, 257–258.
Clinch, P.G. (1979). *N.Z.J. Exp. Agr.* **7**, 407–409.
Clinch, P.G. and Faulke, J. (1977). *N.Z.J. Exp. Agr.* **5**, 185–187.
Clout, G.A. (1956). *Bee Craft*, **38**, 135 only.
Cook, V.A. (1968). *N.Z.J. Agr.* **117**, 61–65.
Crane, E. (1954). *Bee World*, **35**, 29–30.
Crane, E. (1968). *Bee World*, **49**, 113–114.
Crane, E. (1978). *Bee World*, **59**, 164–167.
Dadant, C. and Son (1890). "Langstroth on the Hive and the Honey Bee". Chas. Dadant and Son, Hamilton.
Dade, H.A. (1949). "The Laboratory Diagnosis of Honey-Bee Diseases". Williams and Norgate, London.
Davis, R.E. (1978). *Can. J. Microbiol.* **24**, 954–959.
Deibel, R.H. and Seeley, H.W. (1974). *In* "Bergey's Manual of Determinative Bacteriology", 8th ed. (R.E. Buchanan and N.E. Gibbons, eds.), 490–509. Williams and Wilkins, Baltimore.
De Jong, D. (1976). *Bee World*, **57**, 114–115.
Delfinado, M.D. and Baker, E.W. (1961). *Fieldiana, Zool.* **44**, 53–56.
Dreher, K. (1953). *Z. Bienenforsch.* **121**, 92–97.
Dunham, W.E. and King, P.E. (1934). *J. Econ. Entomol.* **27**, 601–607.
Dzierzon, J. (1882). "Rational Bee-Keeping". Houlston and Sons, London.
Eckert, J.E. (1941). *J. Econ. Entomol.* **34**, 720–723.
Eckert, J.E. (1950). *J. Econ. Entomol.* **43**, 562–564.
Falkener, H.J. (1939). *Bee World*, **20**, 72 only.
Fantham, H.B. and Porter, A. (1912). *Ann. Trop. Med. Parasitol.* **6**, 163–195.
Fantham, H.B. and Porter, A. (1913). *Ann. Trop. Med. Parasitol.* **7**, 569–579.
Fantham, H.B., Porter, A. and Richardson, L.R., (1941). *Parasitology*, **38**, 186–208.
Farrar, C.L. (1947). *J. Econ. Entomol.* **40**, 333–338.
Farrar, C.L. (1954). *Am. Bee J.* **94**, 52–53.
Fekl, W. (1956). *Z. Morphol. Oekol. Tiere*, **44**, 442–458.
Filmer, R.S. (1943). *J. Econ. Entomol.* **36**, 339–341.
Foster, J.W., Hardwick, W.A. and Guirard, B. (1950). *J. Bacteriol.* **59**, 463–470.
Fredskild, B. (1955). *Tidsskr. Biavl*, **89**, 121–123.
Frow, R.W. (1927). *Br. Bee J.* **55**, 437–438.
Furgala, B. (1962). *Glean. Bee Cult.* **90**, 294–295.
Furgala, B. and Gochnauer, T.A. (1969a). *Ann. Bee J.* **109**, 218–219.
Furgala, B. and Gochnauer, T.A. (1969b). *Am. Bee J.* **109**, 380–381.
Fyg, W. (1932). *Schweiz. Bienen-Ztg*, **55**, 1–17.
Fyg, W. (1934). *Landwirt. Jahrb. Schweiz.* **48**, 65–94.
Fyg, W. (1936). *Landwirt. Jahrb. Schweiz.* **50**, 867–880.

Fyg, W. (1939). *Schweiz. Bienen-Ztg*, **9**, 547–552.
Fyg, W. (1948). *Schweiz. Bienen-Ztg*, **12**, 520–529.
Fyg, W. (1954). *Mitt. Schweiz. Entomol. Ges.* **27**, 423–428.
Fyg, W. (1958). *Schweiz. Bienen-Ztg.* **81**, 147–397.
Fyg, W. (1964). *Am. Rev. Entomol.* **9**, 207–224.
Gary, N.D., Nelson, C.I. and Munro, J.A. (1949). *J. Econ. Entomol.* **41**, 661–663.
Giavarini, I. (1937). *Riv. Apicolt.* **1**, 1–7.
Giavarini, I. (1950). *Boll. Zool. Agr. Bachicolt.* **17**, 603–608.
Giavarini, I. (1956). *Ann. Sper. Agr.* **10**, 69–74.
Gibson, T. and Gordon, R.E. (1974). *In* "Bergey's Manual of Determinative Bacteriology" 8th edn, (R.E. Buchanan and N.E. Gibbons, eds.) 529–575. Williams and Wilkins, Baltimore.
Gilliam, M. and Jeter, W.S. (1970). *J. Invertebr. Pathol.* **16**, 69–70.
Giordani, G. (1952). *Ann. Sper. Agr.* **7**, 633–646.
Giordani, G. (1955). *Boll. Ist. Ent. Univ. Bologna* **21**, 61–84.
Giordani, G. (1956). *Ann. Sper. Agr.* **10**, 145–152.
Giordani, G. (1959). *J. Insect. Pathol.* **1**, 245–269.
Glinski, Z. (1972). *Med. Wet.* **28**, 399–405; 458–462; 524–529; 715–722.
Gochnauer, T.A. (1953). *Am. Bee J.* **93**, 410–411.
Gochnauer, T.A. and Corner, J. (1976). *J. Apicult. Res.* **15**, 63–65.
Gochnauer, T.A. and Hamilton, H.A. (1970). *J. Apicult. Res.* **9**, 87–94.
Gochnauer, T.A. and Hughes, S.J. (1976). *Can. Entomologist*, **108**, 985–988.
Gochnauer, T.A. and Margetts, V.J. (1980). *J. Apicult. Res.* **19**, 261–264.
Gochnauer, T.A., Burke, P.W. and Benazet, J. (1979). *J. Apicult. Res.* **18**, 302–308.
Gontarski, H. (1953). *Z. Bienenforsch.* **2**, 7–10.
Gontarski, H. (1954). *Deut. Bienenwirt.* **5**, 162–164.
Greenway, A.R., Greenwood, S.P., Rhenius, V.J. and Simpson, J. (1975) *J. Apicult. Res.* **14**, 63–68.
Gross, K.P. and Ruttner, F. (1970). *Apidologie*, **1**, 401–421.
Gubler, H.U. (1954). *Schweiz. Z. Allg. Pathol.* **17**, 507–513.
Gubler, H.U., Brugger, A., Schneider, H., Gasser, R. and Wyniger, R. (1953). *Schweiz. Bienen-Ztg.* **76**, 268–272.
Guilhon, H. (1950). *Recl Med. Ver.* **126**, 641–660.
Hammer, O. and Karmer, E. (1947). *Schweiz. Bienen-Ztg.* **70**, 190–194.
Harder, A. and Kundert, J. (1951). *Schweiz. Bienen-Ztg.* **74**, 531–544.
Harry, O.G. (1970). *Nature, (London)*, **225**, 964–966.
Haseman, L. (1961). *Am. Bee J.* **101**, 298–299.
Haseman, L. and Childers, L.F. (1944). *Bull. Mo. Agr. Exp. Stn*, **482**, 1–16.
Hassanein, M.H. (1951). *Ann. Appl. Biol.* **38**, 844–846.
Hassanein, M.H. (1952). *Bee World*, **33**, 109–112.
Hayashiya, K., Nishida, J. and Matsubara, F. (1969). *Appl. Entomol. Zool.* **4**, 154–155.
Herrod-Hempsall, W. (1937). "Bee-keeping". *British Bee Journal*, London.
Hertig, M. (1923). *J. Parasitol.* **9**, 109–140.
Hirschfelder, H. (1952). *Z. Bienenforsch.* **1**, 141–170.
Hirschfelder, H. (1957). *Imkerfreund*, **12**, 285–286.
Hirschfelder, H. (1964). *Bull. Apicole*, **7**, 7–17.
Hirst, S. (1921). *Annals and Magazine of Natural History, London*, **7**, 509–519.
Hitchcock, J.D. (1948). *J. Econ. Entomol.* **41**, 854–858.
Hitchcock, J.D. (1936). *J. Econ. Entomol.* **29**, 895–904.
Hitchcock, J.D. (1956). *J. Econ. Entomol.* **49**, 11–14.
Hitchcock, J.D. (1959). *Am. Bee J.* **99**, 418–419.

Hitchcock, J.D. and Wilson, W.T. (1973). *J. Econ. Entomol.* **66**, 901–902.
Hitchcock, J.D., Stoner, A., Wilson, W.T. and Menapace, D.M. (1979). *J. Kans. Entomol. Soc.* **52**, 238–246.
Hoage, T.R. and Peters, D.C. (1969). *J. Econ. Entomol.* **62**, 896–900.
Holst, E.C. (1945). *Science*, **102**, 593–594.
Holst, E.C. (1946). *Glean. Bee Cult.* **74**, 138–139.
Holst, E. and Sturtevant, A.P. (1940). *J. Bacteriol.* **40**, 723–731.
Homann, H. (1933). *Z. Parasitenk.* **6**, 350–415.
Huger, A. (1960). *J. Insect. Pathol.* **2**, 84–105.
Ibragimov, R.P. (1958). *Pchelovodstvo*, **35**, 44–48.
Imms, A.D. (1907). *J. Board Agric. G.B.* **14**, 129–140.
Imms, A.D. (1942). *Parasitology*, **34**, 88–100.
Ingold, G.A. (1950). *Agriculture (London)*, **57**, 35–38.
Ishihara, R. (1969). *J. Invertebr. Pathol.* **14**, 316–320.
Jaekel, S. (1930). *Arch. Bienenkunde*, **11**, 41–92.
Jamieson, C.A. (1955). Progr. Rep. Div. Apic. Can. Dep. Agric. 1949–1953.
Jay, S.C. (1966). *Proc. Entomol. Soc. Manit.* **22**, 61–64.
Jeffree, E.P. (1955). *J. Econ. Entomol.* **48**, 723–726.
Jeffree, E.P. (1959). *Bee World*, **40**, 4–15.
Jeffree, E.P. and Allen, M.D. (1956). *J. Econ. Entomol.* **49**, 831–834.
Johnson, P. (1953). *Nordisk. Bitidskrift* **5**, 98–100.
Jones, D. (1975). *J. Gen. Microbiol.* **87**, 52–96.
Jordan, R. (1937). *Bee World*, **18**, 21 only.
Jordan, R. (1957). *Bienenvater*, **78**, 163–169.
Kaare, H. (1952). *Tidsskr. Biavl.* **86**, 99–101.
Kaare, H. (1953). *Tidsskr. Biavl.* **87**, 69–71.
Kaeser, W. (1952). *Deut. Bienenwirt.* **3**, 21–25.
Kaeser, W. (1954). *Sudwestdeut. Imker*, **6**, 136–138.
Katznelson, H. (1950). *J. Bacteriol.* **59**, 153–155.
Katznelson, H. and Jamieson, C.A. (1952a). *Science*, **115**, 70–71.
Katznelson, H. and Jamieson, C.A. (1952b). *Sci. Agr.* **32**, 219–225.
Katznelson, H. and Jamieson, C.A. (1955). *Can. J. Agr. Sci.* **35**, 189–192.
Katznelson, H. and Lochhead, A.G. (1947). *Sci. Agr.* **27**, 67–71.
Keck, C.B. (1949). *Am. Bee J.* **89**, 514–515.
Kirby, W. and Spence, W. (1826). "An Introduction to Entomology". Longman, London.
Kitaoka, S., Yajima, A. and Azuma, R. (1959). Bull. Nat. Inst. An. Health, No. 37.
Kluge, R. (1963). *Z. Bienenforsch.* **6**, 141–169.
Kramer, J.P. (1960a). *Amer. Midl. Natur.* **64**, 485–487.
Kramer, J.P. (1960b). *J. Insect Pathol.* **2**, 433–436.
Kramer, J.P. (1964). Entomophaga, Memoire Hors Serie No. 2 95–99.
Kramer, V. (1902). *Schweiz. Bienen-Ztg.* **25**, 322–323.
Krebs, H.M. (1957). *Am. Bee J.* **97**, 132–133.
Kshirsagar, K.K. (1966). *Indian Bee J.* **28**, 79–84.
Kulincevic, J.M., Rothenbuhler, W.C. and Stairs, G.R. (1973). *J. Invertebr. Pathol.* **21**, 241–247.
Kulincevic, J.M. and Rothenbuhler, W.C. (1975). *J. Invertebr. Pathol.* **25**, 289–295.
Kulincevic, J.M., Rothenbuhler, W.C. and Stairs, G.R. (1973). *J. Invertebr. Pathol.* **21**, 241–247.
Lack, D. (1954a). "The Natural Regulation of Animal Numbers". Oxford University Press.
Lack, D. (1954b). *In* "The Numbers of Man and Animals". (J.B. Cragg and N.W. Pirie, eds), 47–55. Oliver and Boyd, London.
Laidlaw, H.H. (1979). "Contemporary Queen Rearing". Dadant and Sons, Hamilton, Illinois.

Laigo, F.M. and Morse, R.A. (1968). *Bee World*, **49**, 117–119.
Landerkin, G.B. and Katznelson, H. (1957). *Appl. Microbiol.* **5**, 152–154.
Langridge, D.F. and McGhee, R.B. (1967). *J. Protozool.* **14**, 485–487.
Langstroth. L.L. (1866). A Practical Treatise on the Hive and the Honey Bee. J.B. Lippincott, Philadelphia.
L'Arrivee, J.C.M. (1965). *J. Invertebr. Pathol.* **7**, 408–413.
L'Arrivee, J.C.M. and Hrystak, R. (1964). *J. Insect Pathol.* **6**, 126–127.
Lavie, P. (1950). *Rev. Fr. Apicult.* **12**, 21–24.
Lee, D.C. (1963). *J. Insect Pathol.* **5**, 11–15.
Leuenberger, F. (1933). *Schweiz. Bienen-Ztg.* **56**, 134–147.
Lochhead, A.G.(1942). *J. Bacteriol.* **44**, 185–189.
Lom, J. (1964). *Entomophaga, Memoire Hors Serie, No. 2.* pp. 91–93.
Lotmar, R. (1934), *Bee World*, **15**, 34–36.
Lotmar, R. (1936). *Schweiz. Bienen-Ztg.* **59**, 33–36; 100–104.
Lotmar, R. (1939). *Landwirt. Jahrb. Schweiz.* **53**, 34–70.
Lotmar, R. (1943). *Beih. Schweiz. Bienen-Ztg*, **1**, 261–284.
Lotmar, R. (1946). *Beih. Schweiz. Bienen-Ztg.* **2**, 49–76.
Lotmar, R. (1951). *Z. Vergl. Physiol.* **33**, 195–206.
Lunder, R. (1957). *Nordisk Bitidskrift.* **9**, 107–114.
Mackensen, O. (1951). *Genetics*, **36**, 500–509.
Madelin, M.F. (1960). *Endeavour*, 19, 181–190.
Mages, L. (1956). *Gaz. Apic.* **57**, 86 only.
Malden, W. (1909). *J. Board Agric. G.B.* **15**, 809–825.
Mansi, W. (1958). *Nature (London)*, **181**, 1289 only.
Mathis, M. (1957). *Arch. Inst. Pasteur Tunis*, **34**, 107–113.
Maurizio, A. (1934). Arch. Bienenkunde, **15**, 165–193.
Maurizio, A. (1935). *Ber. Schweiz. Bot. Ges*, **44**, 133–156.
Maurizio, A. (1943). *Schweiz. Bienen-Ztg*, **66**, 376–380.
Maurizio, A. (1945). *Schweiz. Bienen-Ztg*, **67**, 337–369.
Maurizio, A. (1946). *Beih. Schweiz. Bienen-Ztg*, **2**, 1–48.
Melville, R. (1944). *Nature (London)*, **153**, 112 only.
Menapace, D.M. and Wilson, W.T. (1980). *Am. Bee J.* **120**, 761–762.
Milne, P.S. (1942). *Bee World*, **23**, 13–14.
Milne, P.S. (1951). *Agriculture (London)*, **57**, 534–536.
Milne, P.S. (1957). *Bee World*, **38**, 156 only.
Ministry of Agriculture, Fisheries and Food (1956, 1957, 1958, 1959, 1969). Survey of bee health in England and Wales. Ministry of Agriculture London.
Ministry of Agriculture, Fisheries and Food (1974). Beekeeping Statistics. Ministry of Agriculture, Fisheries and Food, London.
Minkov, S.G. and Moiseev, K.V. (1953). *Pchelovodstvo*, **5**, 53–54.
Morgenthaler, O. (1926). *Schweiz. Bienen-Ztg*, **49**, 176–224.
Morgenthaler, O. (1930). *Bee World*, **11**, 49–50.
Morgenthaler, O. (1931a). *Schweiz. Bienen-Ztg*, **53**, 538–545.
Morgenthaler, O. (1931b). *Schweiz. Bienen-Ztg*, **54**, 254–267.
Morgenthaler, O. (1939). *Schweiz. Bienen-Ztg*, **62**, 86–215.
Morgenthaler, O. (1941). *Schweiz. Bienen-Ztg*, **64**, 401–404.
Morgenthaler, O. (1944). *Beih. Schweiz. Bienen-Ztg*, **1**, 285–336.
Morgenthaler, O. (1948). *Bee World*, **29**, 33–34.
Morgenthaler, O. (1963). *Südwestdeut. Imker*, **15**, 102–104.
Morison, G.D. (1931). *Bee World*, **12**, 110–111.

Morison, G.D. (1936). *Rothamsted Conf.* **22**, 17–21.
Morison, G.D., Jeffree, E.P., Murray, L. and Allen, M.D. (1956). *Bull. Entmol. Res.* **46**, 753–759.
Morse, R.A. (1955). *J. Parasitol.* **41**, 553 only.
Mundt, O.J. (1961). *Appl. Microbiol.* **9**, 541–544.
Murray, L. (1952). *Scot. Beekeep.* **28**, 147 only.
Nascimento, C.B. (1971). *Cienc. Abejas.* **1**, 35–39.
Nelson, J.A. (1924). *J. Agr. Res.* **28**, 1167–1214.
Nitschmann, J. (1957). *Deut. Entomol. Z.* **4**, 143–171.
Oertel, E. (1967). *J. Apicult. Res.* **6**, 119–120.
Örösi-Pal, Z. (1934). *Bee World*, **15**, 93–94.
Örösi-Pal, Z. (1935). *Z. Parasitenk.* **7**, 233–267.
Örösi-Pal, Z. (1936). *Z. Parasitenk.* **9**, 125–139.
Örösi-Pal, Z. (1937a). *Z. Parasitenk.* **9**, 125–139.
Örösi-Pal, Z. (1937b). *Zentralbl. Bakteriol. Parasitenk. Infektionskr. Hyg. Abt 2.* **96**, 338–340.
Örösi-Pal, Z. (1938a). *Bee World*, **19**, 64–68.
Örösi-Pal, Z. (1938b). *Zentralbl, Bakteriol. Parasitenk. Infektionskr. Hyg. Abt 2.* **99**, 41–149.
Örösi-Pal, Z. (1963). *J. Apicult. Res.* **2**, 109–111.
Örösi-Pal, Z. (1966). *J. Apicult. Res.* **5**, 27–32.
Otte, E. (1966). *Wien. Tieraerztl. Monatsschr.* **53**, 587–592.
Otte, E. (1967). *Z. Bienenforsch.* **9**, 22–29.
Oudemans, A.C. (1904). *Entomologische berichten, Amsterdam*, **1**, 161–164.
Paillot, A., Kirkor, S. and Granger, A.M. (1944). "L'abeille — anatomie — maladies — enemies". 2nd edn, G. Patissier, Trevoux.
Palmer-Jones, T. (1949). *N.Z.J. Agr.* **79**, 483–486.
Palmer-Jones, T. (1971). *N.Z. Beekeep.* **33**, 6–8.
Palmer-Jones, T. and Robinson, D.S. (1951). *N.Z.J. Sci. Technol. Sect. A.* **32**, 28–38.
Pankiw, P., Bailey, L. Gochnauer, T.A. and Hamilton, H.A. (1970). *J. Apicult. Res.* **9**, 165–168.
Park, O.W. (1935). Iowa Agric. Exp. Stn. J. Paper No. J–327.
Park, O.W. (1953). *Proc. Iowa Acad. Sci.* **60**, 707–715.
Patel, N.G. and Gochnauer, T.A. (1958). *Bee World*, **39**, 36–39.
Patel, N.G. and Gochnauer, T.A. (1959). *Bacterial Proceedings (Soc. Amer. Bacteriologists)*, **59**, 21.
Perepelova, L.P. (1927). *Bee World*, **8**, 133–134.
Phillips, E.F. (1921). U.S. Dep. Agr. Farmers Bull. No, 975.
Phillips, E.F. (1945). Rep. Iowa St. Apiarist, 75–78.
Plurad, S.B. and Hartman, P.A. (1965). *J. Invertebr. Pathol.* **7**, 449–454.
Plus, N., Croizier, G., Duthoit, J.L., David, J., Anxolabehere, D. and Periquet, G. (1975). *C.R.H. Acad. Sci. Ser. D.* **280**, 1501–1504.
Poltev, V.I. (1950). "Bolezni Pchel". Government Publication, Leningrad.
Poltev, V.I. (1953). *Pchelovodstvo*, **5**, 46–48.
Prell, H. (1926). *Arch. Bienenkunde*, **7**, 113–121.
Prell, H. (1927). *Arch. Bienenkunde*, **8**, 1–33.
Reinhardt, J.F. (1942). *Am. Bee J.* **82**, 516 only.
Reinhardt, J.F. (1947). *J. Econ. Entomol.* **40**, 45–48.
Rennie, J. (1923). *North Scot. Agr. Coll. Mem.* **6**.
Rennie, J., White, P.B. and Harvey, E.J. (1921). *Trans. Roy Soc. Edinburgh*, **52**, 737–779.
Ribbands, C.R. (1953). "The Behaviour and Social Life of Honey Bees". Bee Research Association, London.
Rinderer, T.E. and Green, T.J. (1976). *J. Invertebr. Pathol.* **27**, 403–405.
Rinderer, T.E. and Rothenbuhler, W.C. (1969). *J. Invertebr. Pathol.* **13**, 81–86.
Rinderer, T.E., Rothenbuhler, W.C. and Gochnauer, T.A. (1974). *J. Invertebr. Pathol.*, **23**, 347–350.

Rinderer, T.E., Rothenbuhler, W.C. and Kulincevic, J.M. (1975). *J. Invertebr. Pathol.* **25**, 297–300.
Roberts. D. and Smaellie, E. (1958). *N.Z.J., Agric.* **97**, 464–468.
Roeder, K.D. (1953). "Insect Physiology". Chapman and Hall, London.
Roff, C. (1960). *Beekeeping, Queensland*, **3**, 35 only.
Ronna, A. (1936). *Rev. Entomol. (Rio de Janeiro)*, **6**, 1–9.
Root, A.I. (1901). "The A.B.C. of Bee Culture". A.I., Root and Co., Medina.
Rose, R.I. and Briggs, J.D. (1969). *J. Invertebr. Pathol.* **13**, 74–80.
Rothenbuhler, W.C. (1957). *J. Hered.* **48**, 160–168.
Rothenbuhler, W.C. (1958). *Annu. Rev. Entomol.* **3**, 161–180.
Rothenbuhler, W.C. (1964). *Am. Zool.* **4**, 111–123.
Rousseau, M. (1953). *Apiculteur*, **97**, 149–151.
Ruttner, F. and Ritter, W. (1980). *Allg. Deut. Imker-Ztg.* **14**, 129–167.
Ryan, A.F. and Cunningham, D.G. (1950). *Tasmanian J. Agr.* 313–317.
Sachs, H. (1952). *Z. Bienenforsch.* **1**, 148–170.
Sanctuary, C.T. (1924). *Bee World*, **6**, 91–92.
Schneider, H. (1941). *Mitt. Schweiz. Entomol. Ges.* **18**, 318–327.
Schulz-Langner, E. (1958). *Z. Bienenforsch.* **4**, 67–86.
Schulz-Langner, E. (1969). *Z. Bienenforsch.* **9**, 381–389.
Seal, D.W.A. (1967). *N.Z.J. Agr.* **6**, 562 only.
Seltner, B. (1950). *Imkerfreund*, **5**, 56–58.
Showers, R.E., Jones, A. and Moeller, F.E. (1967). *J. Econ. Entomol.* **60**, 774–777.
Simmintzis, G. (1958). *Recl. Med. Vet.* **134**, 919–940.
Simmintzis G. and Fiasson, S. (1949). *C.R.H. Acad. Sci. Ser. D.* **143**, 514–516.
Simmintzis, G. and Fiasson, S. (1951). *Rev. Med. Vet.* **102**, 351–361.
Simpson, J. (1950). *Bee World*, **31**, 41–44.
Simpson, J. (1952). *Bee World*, **33**, 112–117.
Simpson, J. (1958). *Insectes Soc.* **5**, 77–95.
Simpson, J. (1959). *Insectes Soc.* **6**, 85–99.
Singh, S. (1961). *Indian Bee J.* **23**, 46–50
Skou, J.P. (1972). *Friesia*, **10**, 1–24.
Skou, J.P. (1979). *Friesia*, **11**, 265–271.
Skou, J.P. and Holm, S.N. (1980). *J. Apicult. Res.* **19**, 133–143.
Smith, F.G. (1953). *Bee World*, **34**, 233–245.
Smith, F.G. (1960). "Beekeeping in the Tropics". Longmans, London.
Snodgrass, R.E. (1956). "Anatomy of the Honey-Bee". Comstock, Ithaca.
Spiltoir, C.F. (1955). *Amer. J. Bot.* **42**, 501–508.
Spiltoir, C.F. and Olive, L.S. (1955). *Mycologia*, **47**, 238–244.
Steche, W. (1960). *Z. Bienenforsch.* **5**, 49–92.
Steinhaus, E.A. (1949). "Principles of Insect Pathology". McGraw-Hill, New York.
Steinhaus, E.A. (1956). *Hilgardia*, **26**, 107–160.
Stejskal, M. (1958). Int. Congr. Beekeep. Proc. 17th 112 only
Stejskal, M. (1955). *J. Protozool.* **2**, 185–188.
Stejskal, M. (1973). *Die Biene*, **109**, 132–134.
Stevenson, J.H., Needham, P.H. and Walker, J. (1978). Rep. Rothamsted Exp. Sta. 1977, part 2, 55–72.
Sturtevant, A.P. (1919). *J. Econ. Entomol.* **12**, 269–270.
Sturtevant, A.P. (1932). *J. Agric. Res.* **45**, 257–285.
Sturtevant, A.P. (1933). *Am. Bee J.* **73**, 259–261.

Sukhoruka, A.I. (1975). *Pchelovodstvo*, **95**, 18–19.
Sutter, G.R., Rothenbuhler, W.C. and Raun, E.S. (1968). *J. Invertebr. Pathol.* **12**, 25–28.
Tarr, H.L.A. (1937a). *Tabulae Biol.* **14**, 150–185.
Tarr, H.L.A. (1937b). *Ann. Appl. Biol.* **24**, 377–384.
Tarr, H.L.A. (1937c). *Bee World*, **18**, 57–58.
Tarr, H.L.A. (1937d). *Ann. Appl. Biol.* **24**, 369–376.
Tham, V.L. (1978). *Aust. Vet. J.* **54**, 406 only.
Thompson, V.C. (1964). *J. Apicult. Res.* **3**, 25–30.
Thompson, V.C. and Rothenbuhler, W.C. (1957). *J. Econ. Entomol.* **50**, 731–737.
Toumanoff, C. (1951). *Rev. Fr. Apicult.* **68**, 1–325.
Treat, A.E. (1958). *Int. Congr. Ent. Proc. 10th.* **2**, 475–480.
Ulrich, W. (1964). *Z. Bienenforsch*, **7**, 79–86.
Vasliadi, G. (1970). *Pchelovodstvo*, **11**, 16–17.
Vecchi, M.A. (1959). *Ann. Microbiol.* **9**, 73–86.
Virgil (Publius Vergilius Maro). (70–19 B.C.) "The Eclogues and Georgics of Virgil." Trans by J.W. Mackail, (1915). Longmans, Green & Co., London.
Wagner, R.R. (1961). *Bacteriol. Rev.* **25**, 100–110.
Wallace, F.G. (1966). *Exp. Parasitol.* **18**, 124–193.
Wang, Der-I. and Moeller, F.E. (1969). *J. Invertebr. Pathol.* **14**, 135–142.
Wang, Der-I. and Moeller, F.E. (1970a). *J. Econ. Entomol.* **63**, 1539–1541.
Wang, Der-I. and Moeller, F.E. (1970b). *J. Invertebr. Pathol.* **15**, 202–206.
White, G.F. (1906). U.S. Bur. Entomol. Tech. Ser. No. 14.
White, G.F. (1907). U.S. Bur. Entomol. Circ. No. 94.
White, G.F. (1912). U.S. Bur. Entomol. Circ. No. 157.
White, G.F. (1919). U.S. Dep. Agr. Bull. No. 780.
White, G.F. (1920a). U.S. Dep. Agr. Bull. No. 809.
White, G.F. (1920b). U.S. Dep. Agr. Bull. No. 810.
White, P.B. (1921). *J. Pathol. Bacteriol.* **24**, 138–139.
White, J. and Sturtevant, A.P. (1954). *Glean. Bee Cult.* **82**, 658–661.
Wigglesworth, V.B. (1972). "The Principles of Insect Physiology". Chapman and Hall, London.
Wille, H. (1961). *Schweiz. Bienen-Ztg*, **84**, 142–150.
Wille, H. (1967). *Z. Bienenforsch.* **9**, 150–171.
Wilson, C.A. and Ellis, L.L. (1966). *Am. Bee J.* **106**, 131 only.
Wilson, W.T., Elliott, J.R. and Hitchcock, J.D. (1971a). *J. Apicult. Res.* **10**, 143–147.
Wilson, W.T., Elliott, J.R. and Hitchcock, J.D. (1971b). *Am. Bee J.* **111**, 430–431.
Wilson, W.T., Tutt, S.F., Moffett, J.O., Shimanuki, H., Alzubaidy, M.M. and Shipman, C. (1978). *J. Kansas Entomol. Soc.* **51**, 245–252.
Wilson, W.T., Sonnet, P.E. and Stoner, A. (1980). *U.S. Dep. Agr. Handbook.* **335**, 129–140.
Winston, E. (1970). *Am. Bee J.* **110**, 10–11.
Winter, T.S. (1950). *N.Z. Beekeep.* **12**, 16–17.
Woodrow, A.W. (1941a). *Glean Bee Cult.* **69**, 148–151.
Woodrow, A.W. (1941b). *Am. Bee J.* **81**, 363 only.
Woodrow, A.W. (1942). *J. Econ. Entomol.* **35**, 892–895.
Woodrow, A.W. and Holst, E.C. (1942). *J. Econ. Entomol.* **35**, 327–330.
Woodrow, A.W. and States, H.J. (1943). *Am. Bee J.* **81**, 22–23.
Woyke, J. (1962). *J. Apicult. Res.* **1**, 6–13.
Woyke, J. (1963). *J. Apicult. Res.* **2**, 19–24.
Woyke, J. (1965). *J. Apicult,. Res.* **4**, 7–11.
Woyke, J. and Knytel, A. (1966). *J. Apicult. Res.* **5**, 159–164.
Zeitler, H. and Otte, E. (1967). *Zentbl. Vet. Med.* **14**, 186–189.

SUBJECT INDEX

D

E

T

U

V